CHAROPHYTES OF NORTH AMERICA

A Guide to the Species of Charophyta of North America, Central America, and the West Indies

by

R. D. WOOD, Ph.D.

Professor of Botany, University of Rhode Island

Text largely after MONOGRAPH OF THE CHARACEAE

Plates from ICONOGRAPH OF THE CHARACEAE

Available from the Bookstore, Memorial Union
University of Rhode Island
Kingston, Rhode Island 02881 $2.00

STELLA'S PRINTING

West Kingston, Rhode Island

PREFACE

This pocket guide is the realization of the desire by the author to provide a convenient handbook for the general student who is interested in identifying characeae. Comparable data are available in A REVISION OF THE CHARACEAE by Wood and Imahori; but, whereas the REVISION is a highly technical treatise encompassing taxa for the entire world, the present guide includes only enough information to facilitate the study and identification of species and varieties in North America. Most of the text is taken from the MONOGRAPH OF THE CHARACEAE (Vol. I of the REVISION), but certain changes are incorporated in view of progress which has been made in the field since the manuscript was written (1957-59). The plates were taken from the ICONOGRAPH OF THE CHARACEAE (Vol. II of the REVISION) with the permission of the publisher, J. Cramer, Weinheim (now of Lehre), Germany. The writer is indebted to the National Science Foundation for a grant under which much of the original information was compiled. Publication of this guide, however, has been sponsored by the author.

USE OF THIS GUIDE

The purpose of this book is to enable students to identify the species of charophytes of the Western Hemisphere north of the Isthmus of Panama. Technical descriptions are omitted; if needed, they can be found in the author's MONOGRAPH. Illustrations of varieties and forms not included here can be found in the author's ICONOGRAPH. For the most part, identifications can be made by comparing specimens with the pictures; but more advanced students may want to use the KEY in the back of the book. Most species can be identified in the field, especially if diagnostic features are checked with a handlens; but, for confirmation, a compound microscope will be desirable. Varieties are not always easy to identify, and help of a specialist may be needed. If desiring the aid of a specialist, good specimens, either dried or preserved in fluid, together with complete data (collector's name, collection number, date, habitat, precise location), should be sent. Most universities and many museums have botanists who can identify the material or know to whom it should be referred.

The distribution maps provide a useful clue for identifying; but they should be used with caution since the ranges of species are not yet thoroughly known.

A WORD OF CAUTION: charophytes, like many lower plants, exhibit an astounding capacity to produce a wide range of infraspecific variation (both environmental and genetic), and the novice should be cautioned against hasty judgments on perplexing specimens. Even most bizarre specimens will generally conform to one or other of the species described in this book. If in doubt, consult a specialist.

WHAT ARE CHAROPHYTES?

Charophytes belong to the CHARACEAE, a family of plants common in ponds, rivers, and lakes throughout the world. They are not really obscure, as they range in size from less than an inch to more than six feet long; but they are similar to other aquatics and so are easily passed unnoticed. In superficial structure they are similar to specialized algae such as DASYCLADUS, but in gamete and germling formation they resemble bryophytes. They are doubtless a distinct evolutionary line of algae with a long fossil history; and are perhaps somewhat more remote among the algae than are other such unusual groups as the OEDOGONIALES or ZYGNEMATALES.

The plant body has a vertical axis consisting of alternating short nodes and long internodes, and whorls of branchlets occur at the nodes. Rhizoids from the base of the axis grow into the substrate. Remarkably, each internode is but a single large cell, in some cases reaching six inches long and one-eighth inch in diameter. The nodes are more complicated, involving central cells surrounded by numerous peripheral cells. The latter produce the whorls of branchlets and in certain species, especially in the genus CHARA, also produce longitudinal cells which lie flat and cover the internodes with a 1-cell-thick cortex (cortication). The sex organs have a peculiar structure characteristic of this group (examine figures on the plates), and occur at nodes of branchlets and more rarely at the base of branchlets. The true oogonial and antheridial cells are encased in sterile jackets, the female consisting of five spirally twisted cells ("oogonium" or perhaps better, oogoniocarp) and the male of four or eight shieldlike cells ("antheridium" or perhaps better, antheriodiocarp).

Like mosses and most other algae, charophytes have a haploid plant body (thallus); and, as a result, the genetics is considerably unlike that of higher plants. This accounts in part, for instance, for the remarkable uniformity often found among individuals of an entire pond.

THE NAME

CHAROPHYTE means "Chara plant." It has two uses; first, as a technical name for a division of the plant kingdom including all chara-like plants both fossil and modern, and, second, as a common name for present-day chara-like plants of the family CHARACEAE. It is used in the second sense in this book. These plants are sometimes referred to as muskgrasses or stoneworts, but in our country the genus names CHARA, NITELLA, and TOLYPELLA are usually used.

WHERE THEY OCCUR

Charophytes occur in fresh-water and brackish marshes, ponds, lakes, and rivers. Nitellas predominate in mildly acid water as in igneous rock areas, charas predominate in hard water, while tolypellas are nearly restricted to hard water of limestone areas. They occur in shallow waters, but are most common between 6 and 25 feet deep. In unusual circumstances, they are reported to grow at depths of 60 and 90 feet, and one highly modified taxon was dredged up alive from 400 feet deep in Lake Tahoe, California.

HOW MANY SPECIES

There are 81 distinct species of charophytes in the world by a recent count—22 in North America; but there are so many variations that thousands of different kinds have been described.

VARIATIONS

Even within a single clone of charophytes, the different shoots may vary with conditions. In quiet water with dim light, plants tend to be longer, slenderer, with reduced appendages and fewer gametangia; while in shallow water with bright light and wave action, the plants tend to be shorter, stouter, with longer appendages and more gametangia. Deep-water forms may remain vegetative throughout the season. In unusual conditions, shoots are small, retaining features of young shoots but producing fertile gametangia. Such stages are called "juveniles."

ECONOMIC VALUE

The importance of charophytes in North America

is indirect—as food for migratory waterfowl, protection of fish fry, and as a nuisance in shallow waters of reservoirs and recreational areas. In other countries, they have been used variously for sulfur baths, cattle food, fertilizers, scouring and filtering agents, and even for supposed control of mosquito larvae.

COLLECTING, MOUNTING, PRESERVING, AND LABELING

Charophytes are easily collected by hand or rake, and in deeper water by skin diving or dredging with a grapnel-like plant hook. They must be kept moist until mounted, as in a plastic bag or in moistened newspaper. Keep careful notes on date and place of collection.

Always mount specimens of good material, and preserve a duplicate sample in fluid. For mounting, use herbarium paper or high quality bond paper. Sort out specimens in a pan of water, sink, or bathtub; and slide paper under selected specimen, gently raising it from the water. Drain until paper loses its gloss, then put it in a fold of newspaper and cover the specimen with a sheet of waxed paper. They can be dried in a plant press at moderate heat, or by placing blotters or several layers of dry newspaper between the newspaper folds with samples and changing them daily. When dry, specimens will lose the cool feel and fresh smell; and then the waxed paper can be cautiously peeled off. Fragile specimens are best placed in a packet, and then mounted on an herbarium sheet. Enter data on each sheet.

For preserving, cover a moderate amount of material in a jar with 4% formalin (4 parts formalin and 96 parts water). For security, be sure solution is alkaline (can be tested with litmus paper), and add borax if necessary; add a little glycerine to prevent damage to specimens in case the fluid dries out. Data can be written on slips of herbarium paper with pencil or India ink, and slid around the glass inside the bottle.

Labeling is most important, and should include collector's name and collection number, date, habitat, and precise location. The identification can be added, or changed, later.

CHARACEAE
Herbarium of R. D. Wood
University of Rhode Island

No. 67-10-5-3 Chara vulgaris L.
var. gymnophylla (A.Br.) Nym.

In 6 ft. of clear water, gravel-mud bottom, NE bay just E of Gansa, Lake Tanganyika, Tanganyika.

Coll.: R. D. Wood Oct. / 5 /19 67

Det.: R.D.W. Dec. / 10 /19 67

INFORMATION IS NEEDED

You can help establish the true range of species of charophytes by collecting intensively in your area. Your findings, together with findings of others, can then be assembled to advance a thorough knowledge of the distribution of these plants in your county, state, or province. Such information would be a valuable contribution.

Important information on what conditions each species needs can be obtained by careful field observation; and, if you have access to certain additional equipment, making careful measurements of acidity, temperature, minerals, dissolved gases, etc., will extend your knowledge. Perhaps you can find what conditions kill out various species, and where the charophytes occur in aquatic succession.

Sift the soil in your area and examine under the microscope. You might be walking on fossilized oospores of charophytes. They have been found in several states.

TAXONOMY

Identification of charophytes has always been considered difficult, but part of the problem is eliminated in the present book by using broad, well-defined species. Within the limits of certain of these species, however, are genetically different kinds of more or less distinctness, and with smaller and smaller differences. These many different strains create a problem, for they are usually so intermixed that no practical classification seems capable of handling them. On the other hand, individual strains occasionally become established in a pond or area; and, although really not distinct species, get described as such. Some workers become fascinated by these minor variations, and insist on recognizing them as species; whereas, others would consider them simply as forms. If they are to be considered as species, they are certainly small ones, a lot smaller than the broad species; and they might best be designated as microspecies. A list of microspecies appears in the author's MONOGRAPH (Wood, 1965: 763); and, in the present guide, such names are marked "microsp." in the list of synonyms on p. 60. One frontier of research involves establishing the degree of cross fertility of these microspecies. However, the reader is advised that recognition or naming of forms, or microspecies if you will, is a complex endeavor best left to advanced researchers.

GENERA OF THE REGION

CHARA: fig. 1. Note the simple (undivided) branchlets. Also, see figs. 7-9, 14, 16-18.

NITELLA: fig. 5. Note the forked (or furcate) branchlets, a feature which is present but obscure in fig. 4; the rays are reduced to a tiny terminal corona barely visible to the naked eye. Also, see figs. 2-6, 10-12.

TOLYPELLA: figs. 13 and 15. Note the coarse, bird's-nestlike heads, and the monopodial structure of branchlets; i. e., having a central axis with multicellular lateral rays and end segment.

LEGEND AND GUIDE TO TERMINOLOGY

(See page 10 for list of terms)

1. Habit; with main axis and whorls of simple (undivided) branchlets at the nodes (separated by internodes) (C. vulgaris).
2. 1-furcate branchlet (N. flexilis).
3. 1-celled dactyl; anarthrodactylous (N. flexilis).
4. Habit; axillary fertile heads; branchlets appear simple but are obscurely 1-furcate, the dactyls being very abbreviated (brachydactylous) (N. translucens).
5. Habit; 3—4-furcate branchlets, and a lateral branch (N. gracilis).
6. Fertile branchlet, 3—4-furcate (forked, divided); gametangia at branchlet furcations, the antheridia (round) terminal at each ray, the oogonia lateral; the lowest branchlet furcation is fertile (N. gracilis).
7. Ecorticate branchlet, bract-cells at nodes; bracteoles attached just below oogonia (2 shown at the fertile nodes), oogonium and antheridium at same node (conjoined) (C. braunii).
8. Corticated branchlet (phloeopodous) (C. globularis).
9. Basal branchlet segment naked (gymnopodous) (C. zeylanica).
10. 2-celled dactyl (arthrodactylous) with reduced end-cell (N. gracilis).
11. 3-celled dactyl (N. furcata).
12. More-than-3-celled dactyl (polyarthrodactylous).
13. Habit; note simple sterile branchlets, while the fertile branchlets are compacted into coarse bird's-nestlike heads (T. nidifica).
14. Ecorticate axis; stipulodes in 1 tier (haplostephanous), alternate (C. braunii).
15. Monopodial branchlet, with a central axis having a ray-bearing node; the end segment and lateral rays are multicellular (T. nidifica).
16. Corticated axis, 1-cort. (haplostichous); stipulodes in 1 tier (haplostephanous) (C. canescens).
17. Corticated axis, 2-cort. (diplostichous), with the primary cells more prominent (tylacanthous); stipulodes in 2 tiers (diplostephanous) (C. vulgaris).
18. Corticated axis, 3-cort. (triplostichous); stipulodes tiny but in 2 tiers (diplostephanous) (C. globularis).

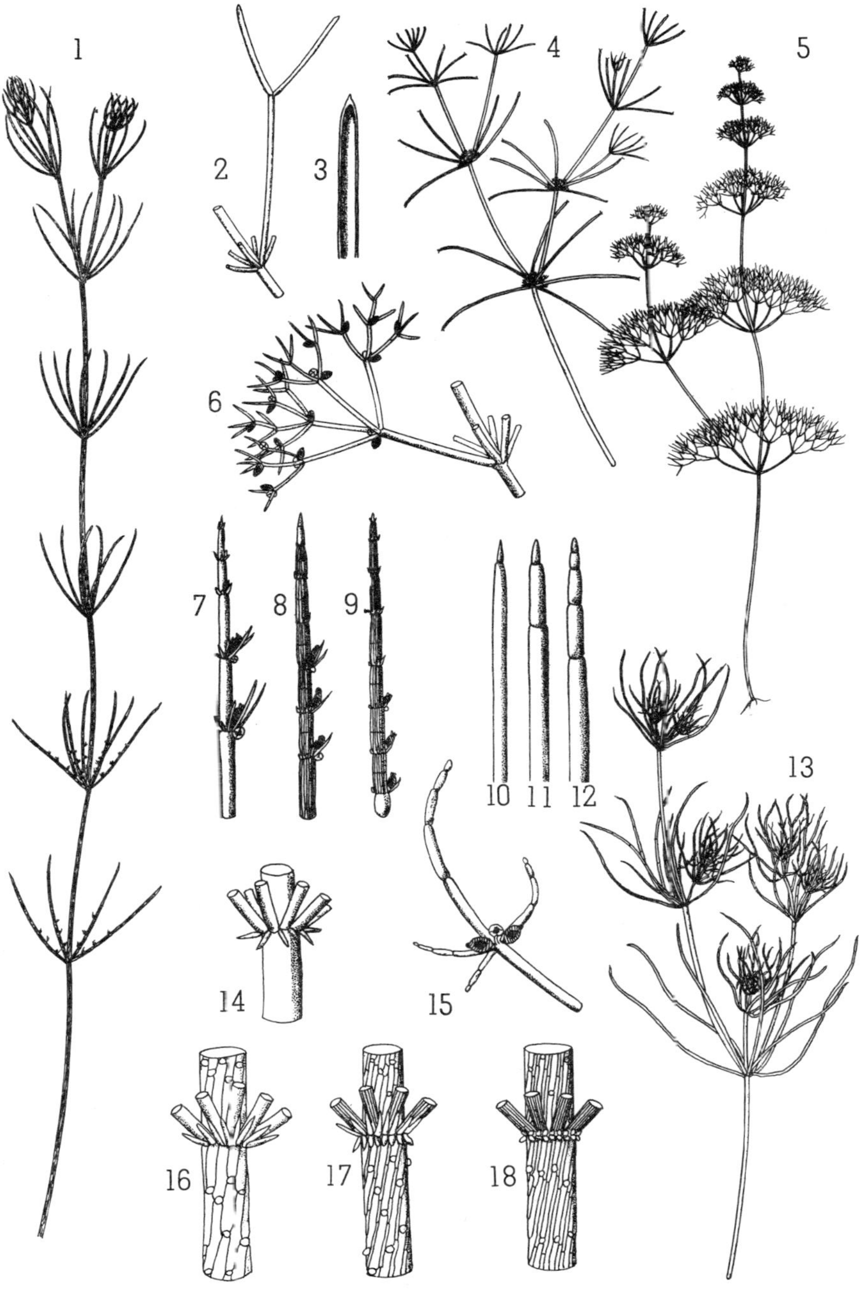
1
2
3
4
5
6
7
8
9
10
11
12
13
14
15
16
17
18

LIST OF TERMS

(Refer to figures and legend on pages 8 and 9.)

ANNOTATED CLASSIFICATION FOR CHAROPHYTES OF NORTH AMERICA

(For meanings of terms, see p. 10)

Family CHARACEAE

Tribe CHAREAE (coronula 5-celled)

CHARA (branchlets simple; stipulodes present)
subgen. CHARA (diplostephanous)
sect. CHARA (diplostichous)..... 1. Chara vulgaris
sect. DESVAUXIA (haplostichous)..2. C. canescens
sect. GROVESIA (triplostichous)
subsect. GROVESIA (phloeopodous)
.... 3. C. globularis
subsect. WILLDENOWIA (gymnopodous)
.... 4. C. zeylanica
subgen. CHAROPSIS (haplostephanous)
sect. CHAROPSIS (ecorticate)....... 5. C. braunii
sect. AGARDHIA (corticate)
subsect. AGARDHIA (monoecious; delicate)
.... 6. C. fibrosa
subsect. WALLMANIA (dioecious; robust)
.... 7. C. hornemannii
.... 8. C. buckellii

Tribe NITELLEAE (coronula 10-celled)

NITELLA (branchlets furcate; with dactyls)
subgen. NITELLA (anarthrodactylous)
sect. NITELLA (dactyls not acuminate)
.... 9. Nitella flexilis
.... 10. N. macounii
sect. RAJIA (dactyls acuminate)..11. N. acuminata
.... 12. N. bastinii
sect. BROWNIA (heteroclemous) ... 13. N. clavata
sect. RIDDELLIA (heads tiny, with mucus; brachydactylous) .... 14. N. cernua
subgen. HYELLA (arthrodactylous; end-cells obtuse, not reduced) .... 15. N. capillata
subgen. TIEFFALLENIA (arthrodactylous; end-cells acute, reduced)
sect. TIEFFALLENIA (mucronate)...16. N. furcata
sect. GIOALLENIA (not mucronate)
.... 17. N. tenuissima
.... 18. N. gracilis
sect. DECANDOLLEA (heteroclemous)
.... 19. N. hyalina
sect. PERSOONIA (heads tiny, without mucus)
.... 20. N. translucens

TOLYPELLA (branchlets monopodial; rays 3–more-celled)
sect. TOLYPELLA (end-cells obtuse)
.... 21. Tolypella nidifica
sect. ROTHIA (end-cells acute)... 22. T. intricata

1. Chara vulgaris L.

(Chara contraria, C. foetida, etc., see p. 60)

FIELD CLUES: common chara of hard-water lakes, with smooth appearance but with stony lime incrustation, stipulodes in 2 tiers, cortication throughout, medium large (to 18").

DIAGNOSTIC FEATURES:
- monoecious
- stipulodes in 2 tiers, usually small and blunt
- spine-cells absent or small, rarely geminate
- cortex on axes and branchlets; axis diplostichous

HABITAT: forming dense "forests" or carpets in shallow water. Clumps more isolated on wave-swept beaches. Hard-water lakes and rivers.

VARIATIONS: considerable ecological and great genetic variation. Some unusual traits which occur are
- geminate spine-cells
- rare bifid bract-cells and bracteoles
- huge and irregular stipulodes
- elongate spine-cells, bract-cells, or bracteoles
- very slight lime
- irregular or few corticated branchlet segments
- incomplete cortex
- enlarged primary (tylacanthous) or secondary (aulacanthous) cortical cells

CLASSIFICATION: var. VULGARIS, the common variety with normal cortex.
var. INCONNEXA (T. F. A.) R. D. W., incomplete and variable cortex.

NOTE: many strains of var. vulgaris occur but separation is impractical and such attempts are best left for advanced research.

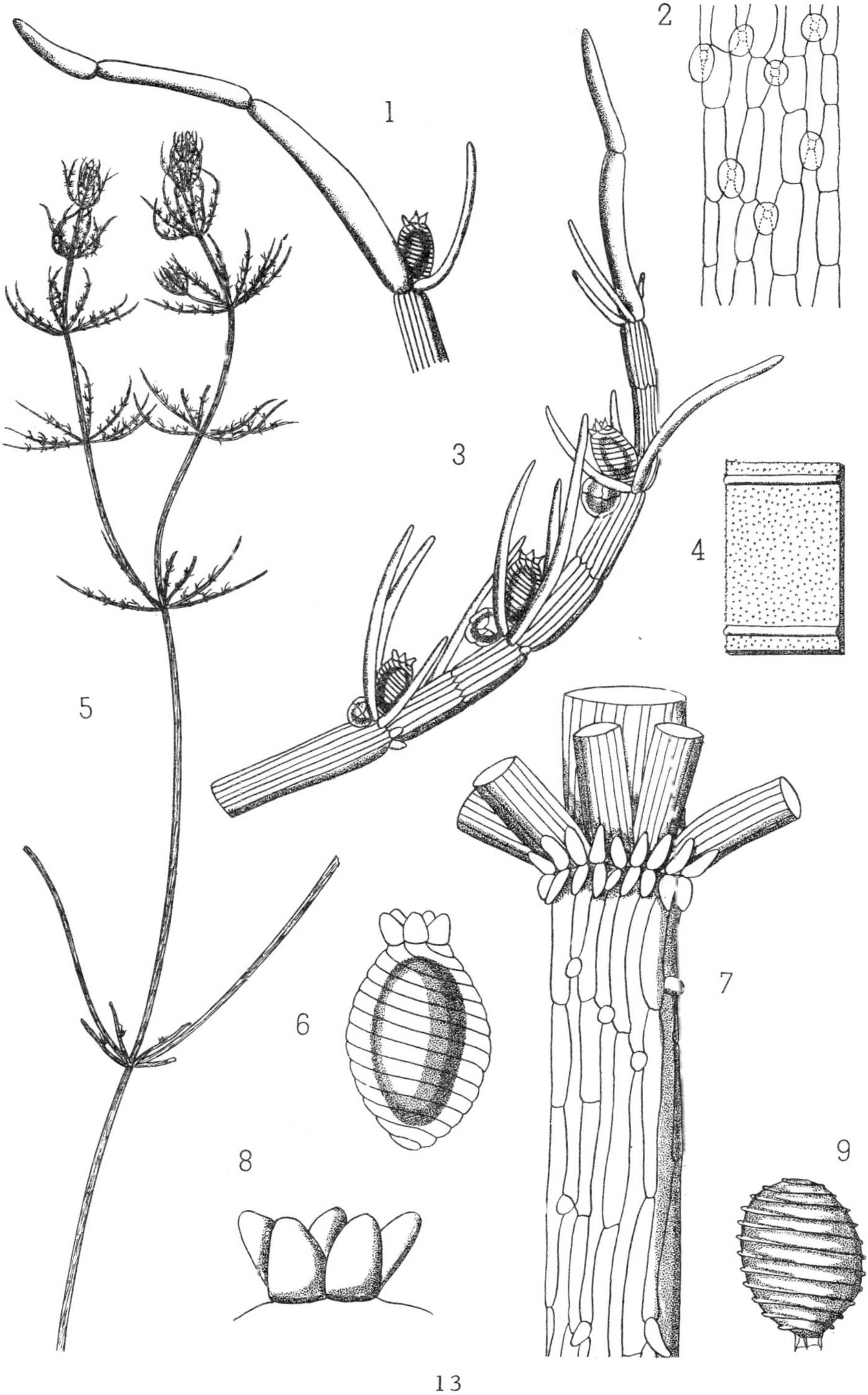
1
2
3
4
5
6
7
8
9

2. Chara canescens Desv. & Lois. in Lois.

(Chara evoluta, C. hirsuta)

FIELD CLUES: small spiny-looking chara of saline or brackish water; commonly smaller than 8".

DIAGNOSTIC FEATURES:
dioecious or monoecious
stipulodes in 2 tiers, long and pointed
spine-cells clustered, 2—5
cortex on axes and branchlets; axis haplostichous (or somewhat overlapping)

HABITAT: forming dense carpets in shallow water. Brackish or saline pools and lagoons.

VARIATIONS: fairly uniform, though extremes reach 2 ft. long (in Calif. and Neb.). Dioecious and monoecious strains.

CLASSIFICATION: not divided into varieties.
NOTE: usually considered dioecious (female), but the monoecious strain is widespread in North America while the dioecious strain is mostly restricted to coastal regions (New Engl., Brit. Col.).

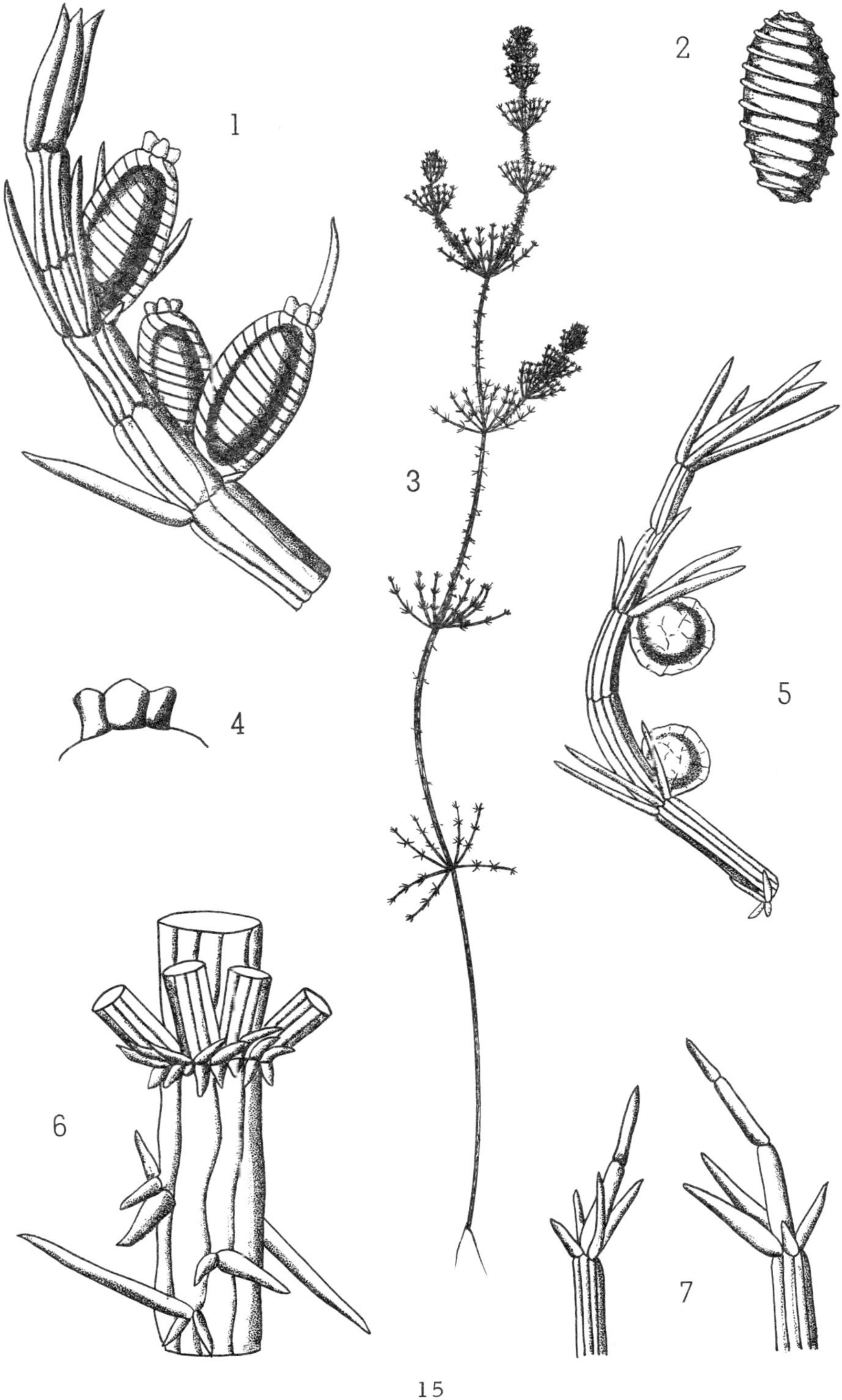
1
2
3
4
5
6
7

3. **Chara globularis** Thuill.

(Chara aspera, C. delicatula, C. fragilis, etc., see p. 60)

FIELD CLUES: common chara of both hard- and soft-water lakes, with very smooth appearance and only moderate lime incrustation, stipulodes tiny or obscure, cortication throughout, medium large (to 18").

DIAGNOSTIC FEATURES:
- monoecious; rarely dioecious
- stipulodes in 2 tiers, occasionally appearing in 1 tier, or obscure to entirely absent
- spine-cells absent or obscure, but may be well developed and in clusters (var. aspera)
- cortex on axes and branchlets; axis triplostichous

HABITAT: forming more or less dense "forests" in shallow water of both hard- and soft-water lakes, ponds, rivers, and coastal ponds. Brackish water examples (often dioecious) commonly have larger stipulodes and larger spine-cells.

VARIATIONS: considerable ecological and also much genetic variation. Some unusual traits which occur are
- enlarged stipulodes
- stipulodes in 1 or 2 tiers
- enlarged primary cortical cells (tylacanthous)
- dioecious or subdioecious
- spine-cells enlarged and in clusters.

CLASSIFICATION: var. GLOBULARIS, the common smooth form with normal cortex and tiny stipulodes (may be absent).

var. VIRGATA (Kütz.) R. D. W., stipulodes in 1 tier.

var. LEPTOSPERMA (A. Br.) R. D. W., stipulodes in 2 tiers, well developed.

var. ASPERA (Deth. ex Willd.) R. D. W., spine-cells well developed, clustered; invariably dioecious.

NOTE: a nearly dioecious strain of var. virgata occurs in central North America.

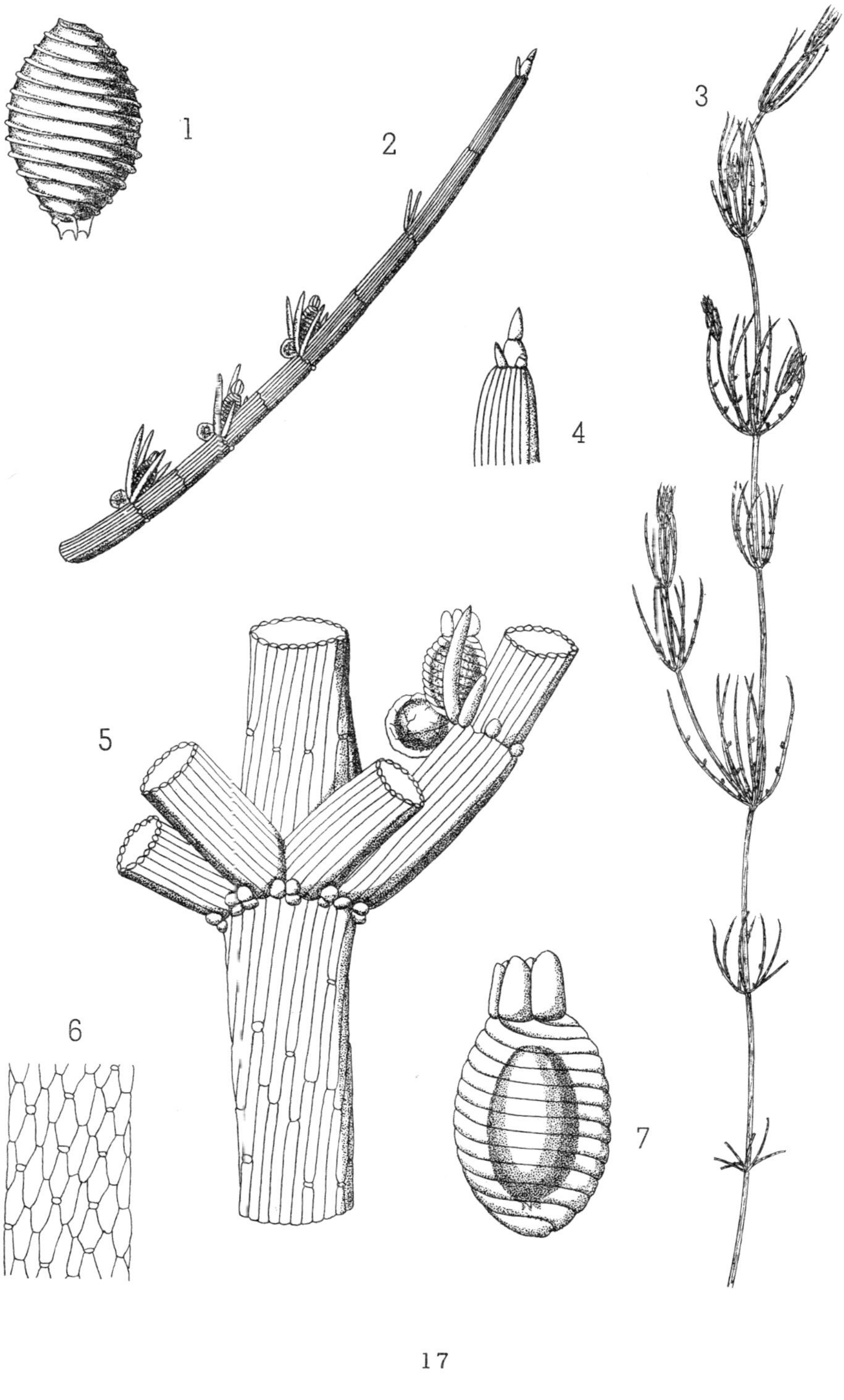
1
2
3
4
5
6
7

4. Chara zeylanica Kl. ex Willd.

(Chara gymnopus, C. sejuncta, etc., see p. 60)

FIELD CLUES: fairly common chara of much of region, highly variable, but with distinct stipulodes in 2 tiers and with basal branchlet segment uncorticated, often robust (to 2').

DIAGNOSTIC FEATURES:
- monoecious; rarely dioecious
- stipulodes in 2 well-developed tiers
- spine-cells generally distinct, solitary
- cortex on axes and branchlets, axis triplostichous; occasionally absent from branchlets, and rarely entirely absent
- gametangia frequently isolated (sejoined)

HABITAT: in isolated clumps or in thin "forests" in shallow to deeper water, soft- and hard-water lakes, ponds, and rivers, and in brackish ponds and lagoons.

VARIATIONS: considerable ecological and extreme genetic variation. Some unusual traits which occur are
- sejoined gametangia
- elongated or shortened spine-cells, bract-cells, and bracteoles
- elongated basal branchlet segment
- elongated branchlet end segment
- modification of bract-cells of lowest branchlet node
- reduced number of branchlet segments
- reduced and irregular number of corticated branchlet segments
- dioecious

(continued on page 20)

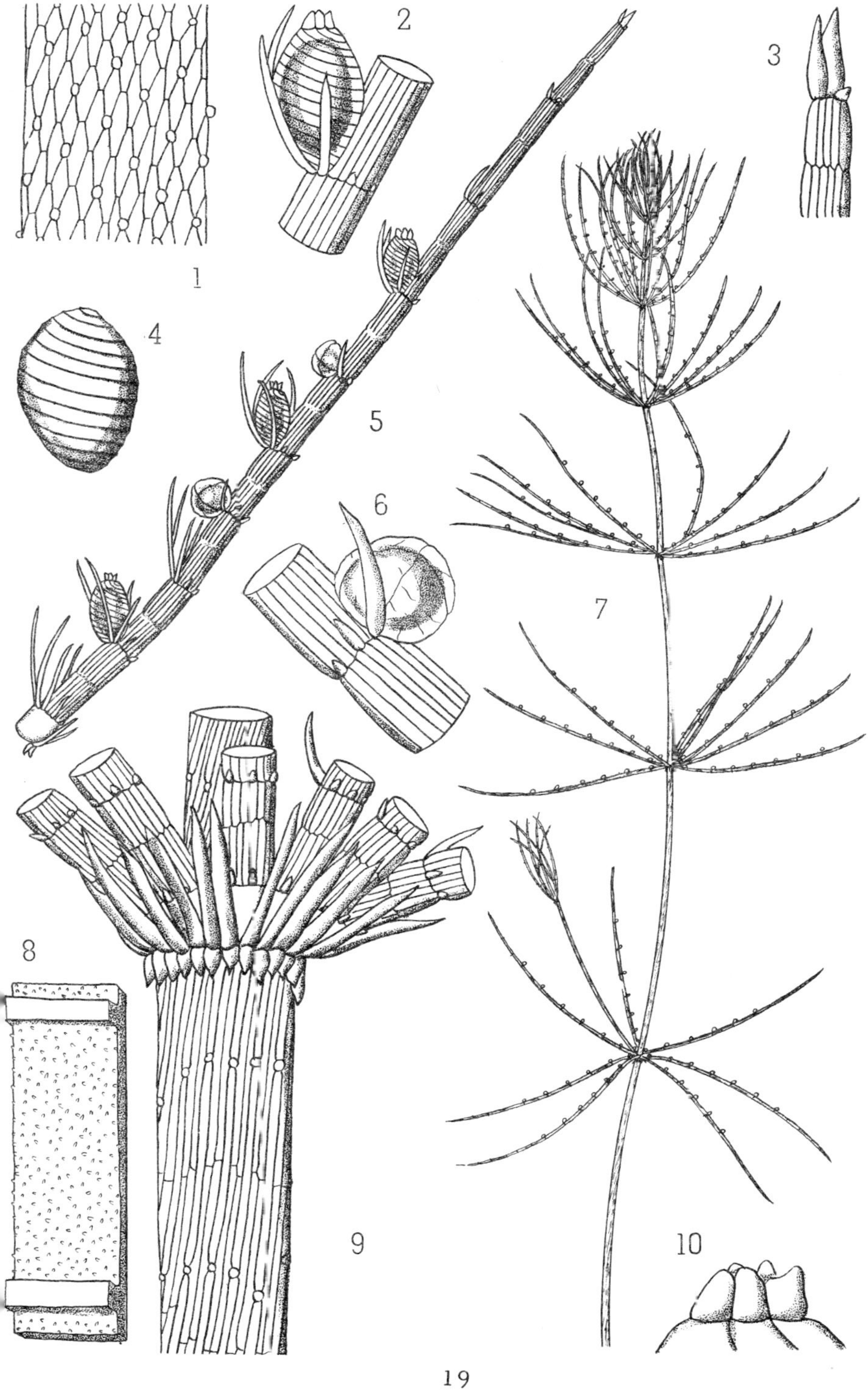
1
2
3
4
5
6
7
8
9
10

4. Chara zeylanica (continued)

CLASSIFICATION: var. ZEYLANICA, the common form with complete cortex and fairly robust stature.

var. SEJUNCTA (A. Br.) R. D. W., small and slender, with gametangia consistently isolated.

var. DIAPHANA (Meyen) R. D. W., branchlet cortication irregular or absent.

var. BRITTONII (T. F. A. ex Rob.) R. D. W., totally without cortex.

NOTE: dioecious strain (male only) of Gatun Lake, Panama, (f. kenoyeri) is extremely spinescent; while dioecious strain of South America is typical of var. zeylanica.

Plate on page 21

Chara zeylanica f. kenoyeri (Howe) R. D. W.

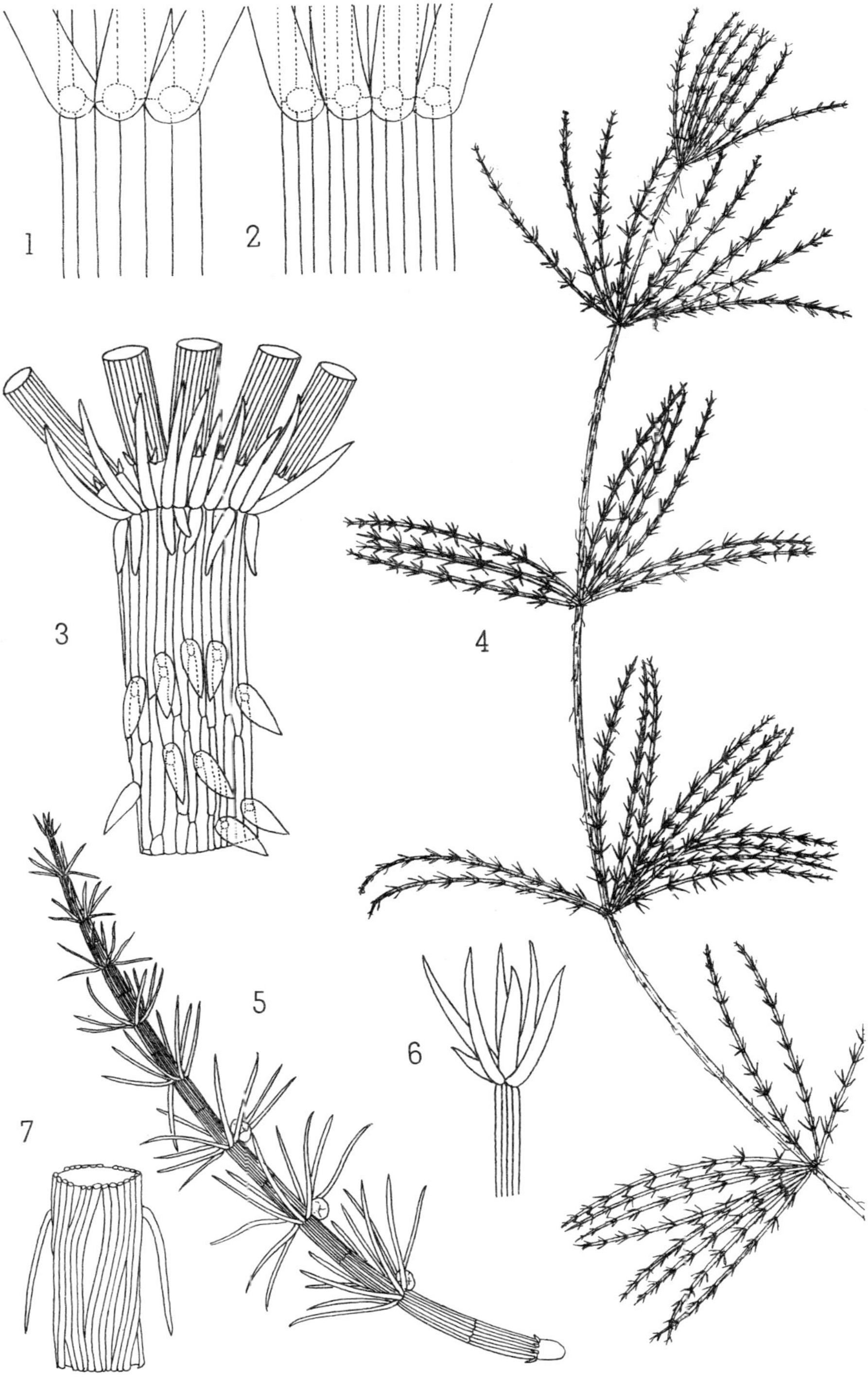
1
2
3
4
5
6
7

5. Chara braunii Gm.

(Chara coronata, C. schweinitzii, etc., see p. 61)

FIELD CLUES: the only totally ecorticated species with undivided branchlets; occasionally ring-banded incrustation, medium large (to 15").

DIAGNOSTIC FEATURES:
monoecious
stipulodes in 1 tier, alternating with branchlets
spine-cells none
cortex none
branchlets with tiny terminal corona

HABITAT: in isolated clumps in shallow to deeper water of both hard- and soft-water lakes and ponds. Occasional in slightly brackish water of coastal ponds.

VARIATIONS: considerable ecological, but little genetic variation. Some unusual traits which occur are
- elongation of stipulodes and bract-cells
- reduction in spore size
- reduction in length of some branchlet segments

CLASSIFICATION: not divided into varieties.

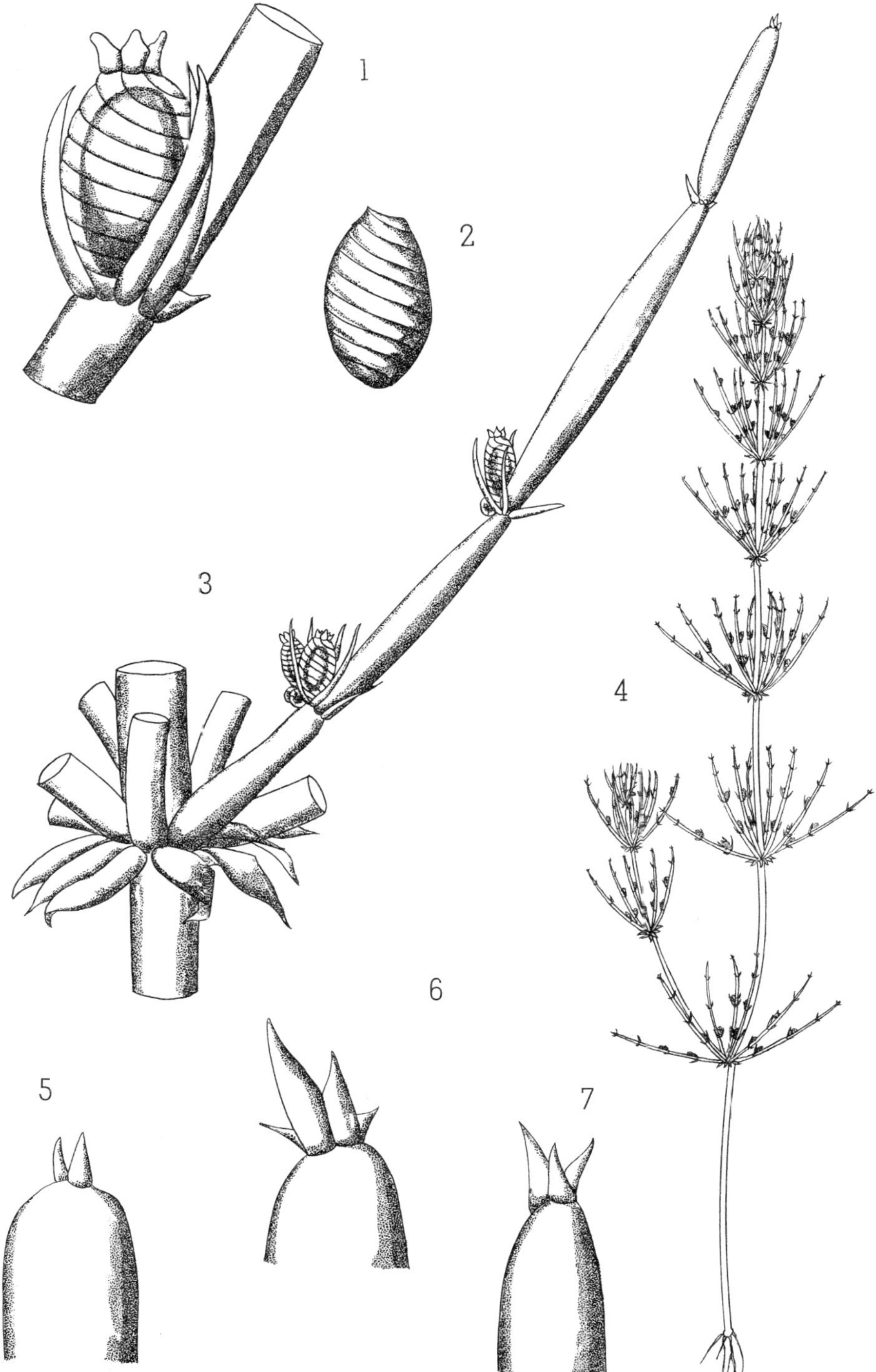
1
2
3
4
5
6
7

6. Chara fibrosa Ag. ex Bruz.

(Chara gymnopitys, C. hydropitys, etc., see p. 61)

FIELD CLUES: slender, delicate chara; with large bract-cells; the only small species with axial cortex and stipulodes in 1 tier; small (to 12").

DIAGNOSTIC FEATURES:
- monoecious
- stipulodes in 1 tier, 1—2 per branchlet
- spine-cells small to large, solitary
- cortex on axis, partial or absent from branchlets, but basal branchlet segment invariably naked; axis 2—3-corticate

HABITAT: isolated clumps in shallow water of soft- and hard-water lakes, ponds, and rivers.

VARIATIONS: considerable ecological and genetic variation. Some unusual traits which occur are
- extreme smallness
- heavy incrustation
- elongate stipulodes, bract-cells, and bracteoles
- reduction in number of corticated branchlet segments
- length of basal branchlet segment
- spine-cells rarely clustered

CLASSIFICATION: var. FIBROSA, normal form with naked branchlets.
- var. HYDROPITYS (Reich.) R. D. W., at least some of branchlet segments corticated.
- var. MYRIOPHYLLA (Müll. ex A. Br.) R. D. W., naked branchlets, but dense spine-cells occurring in clusters (only from Canadaigua, N. Y.).

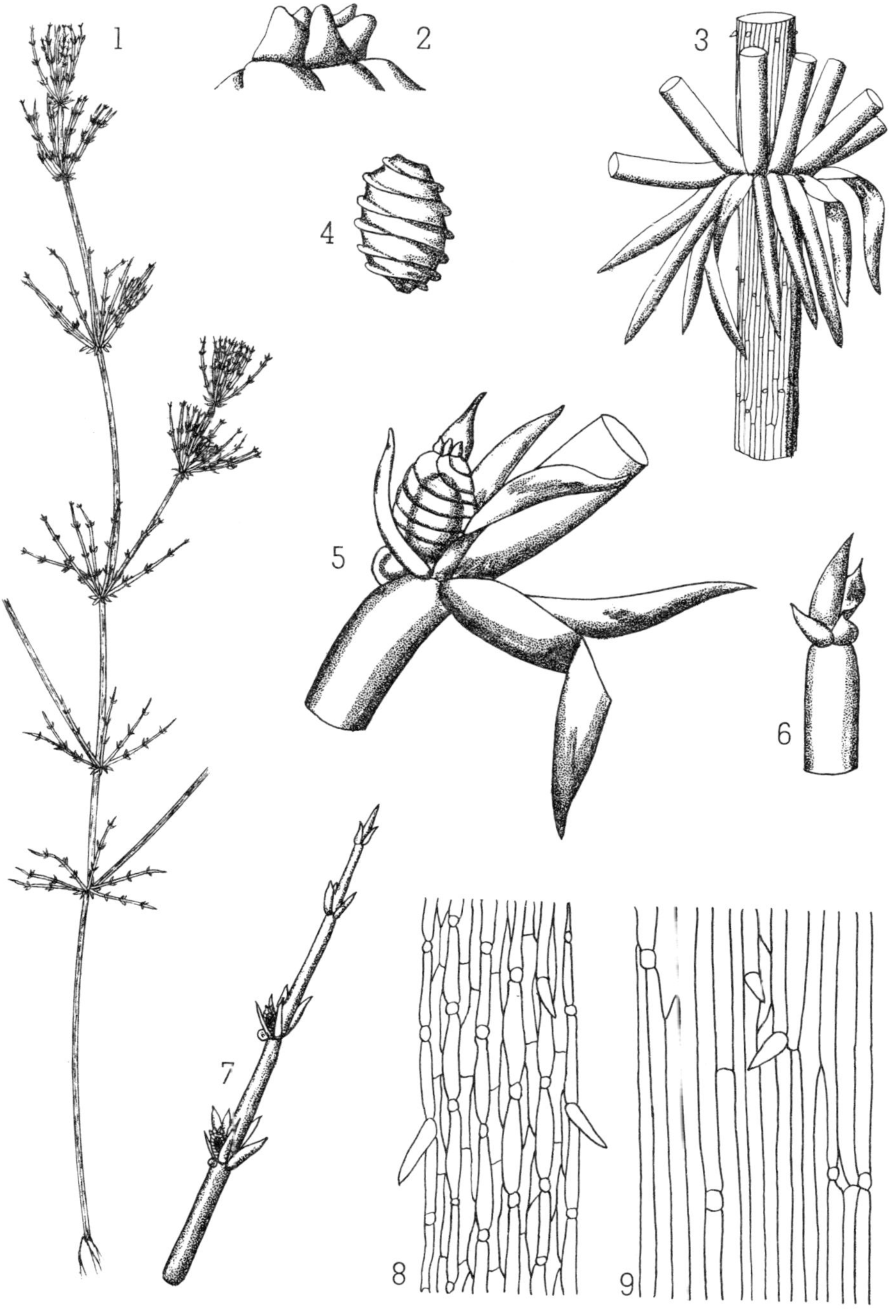
1
2
3
4
5
6
7
8
9

7. Chara hornemannii Wallm.

(Chara longifolia, C. nordhoffiae)

FIELD CLUES: large, spinescent-appearing chara of brackish water, with axial cortex and naked branchlets; robust (to 3').

DIAGNOSTIC FEATURES:
- dioecious
- stipulodes in 1 tier, 1—2 per branchlet
- spine-cells generally elongate, solitary
- cortex only on axis, branchlets naked; axis diplostichous

HABITAT: shallow brackish or saline ponds and lagoons.

VARIATIONS: considerable ecological and some genetic variation. Also, juveniles occur which are small, naked, yet fertile. Some unusual traits which occur are
- reduced size and number of spine-cells, stipulodes, and bract-cells
- bending down of some bract-cells
- reduction in axial cortex, sometimes largely absent
- accumulation of lime at nodes or bulbils

CLASSIFICATION: not divided into varieties.

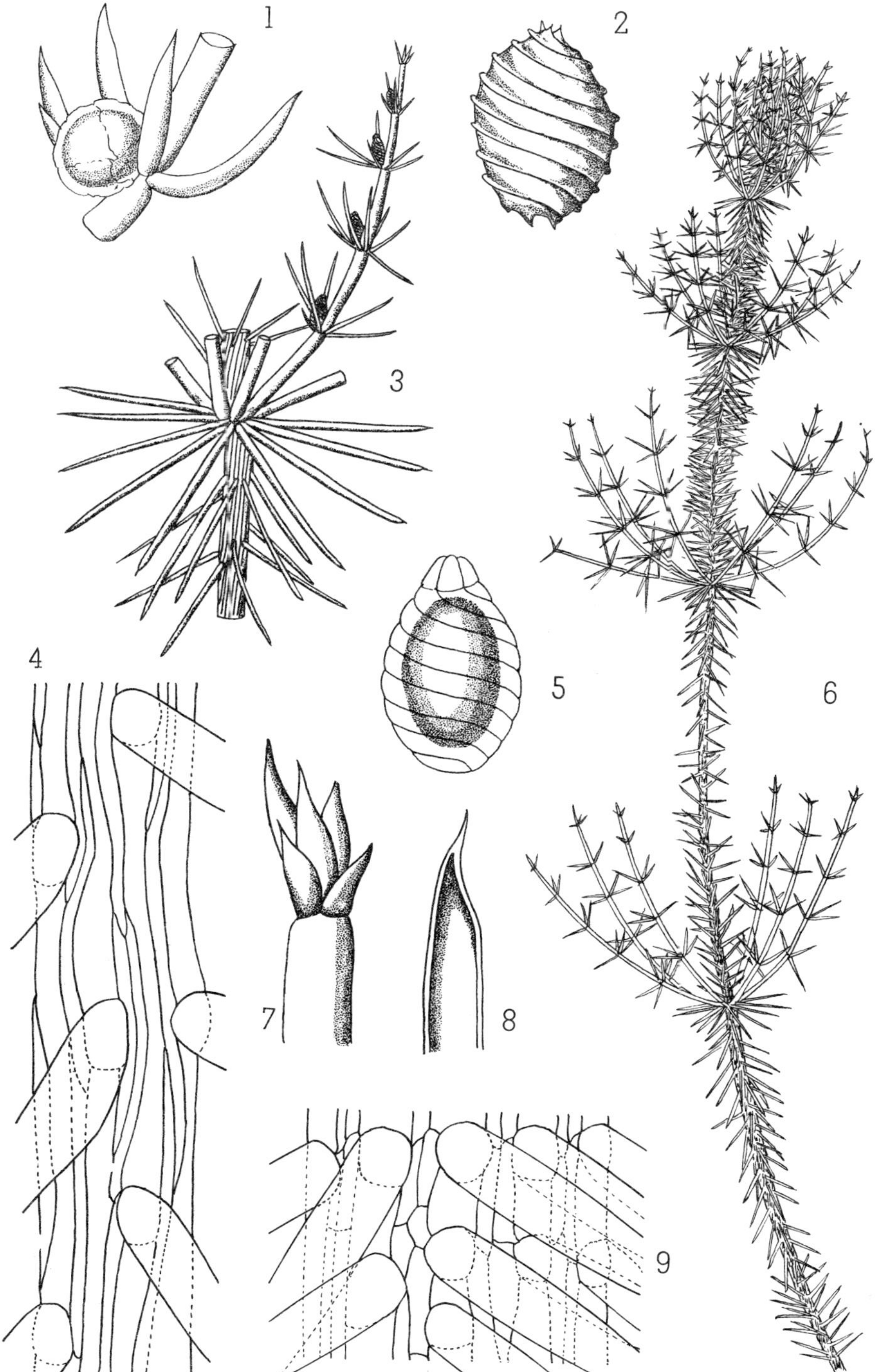
1
2
3
4
5
6
7
8
9

8. **Chara buckellii** G. O. A.

FIELD CLUES: stout, spinescent-appearing chara of brackish water, with axial cortex, naked branchlets, stipulodes stout and in 1 tier; medium size (to 18").

DIAGNOSTIC FEATURES:
- dioecious
- stipulodes in 1 tier, one per branchlet, and inserted directly below the branchlet (i.e., opposite)
- spine-cells obscure or absent
- cortex only on axes, branchlets naked; axis diplostichous

HABITAT: shallow water of brackish pond.

VARIATIONS: not well known.

CLASSIFICATION: not divided into varieties.

NOTE: hybridization studies suggest that this is conspecific with a weak form of C. hornemannii from New Mexico as well as with Nitellopsis bulbillifera from Argentina. Further data are needed for re-evaluation of the species.

NOTE: Nitellopsis obtusa, reported from Guadeloupe, might belong here.

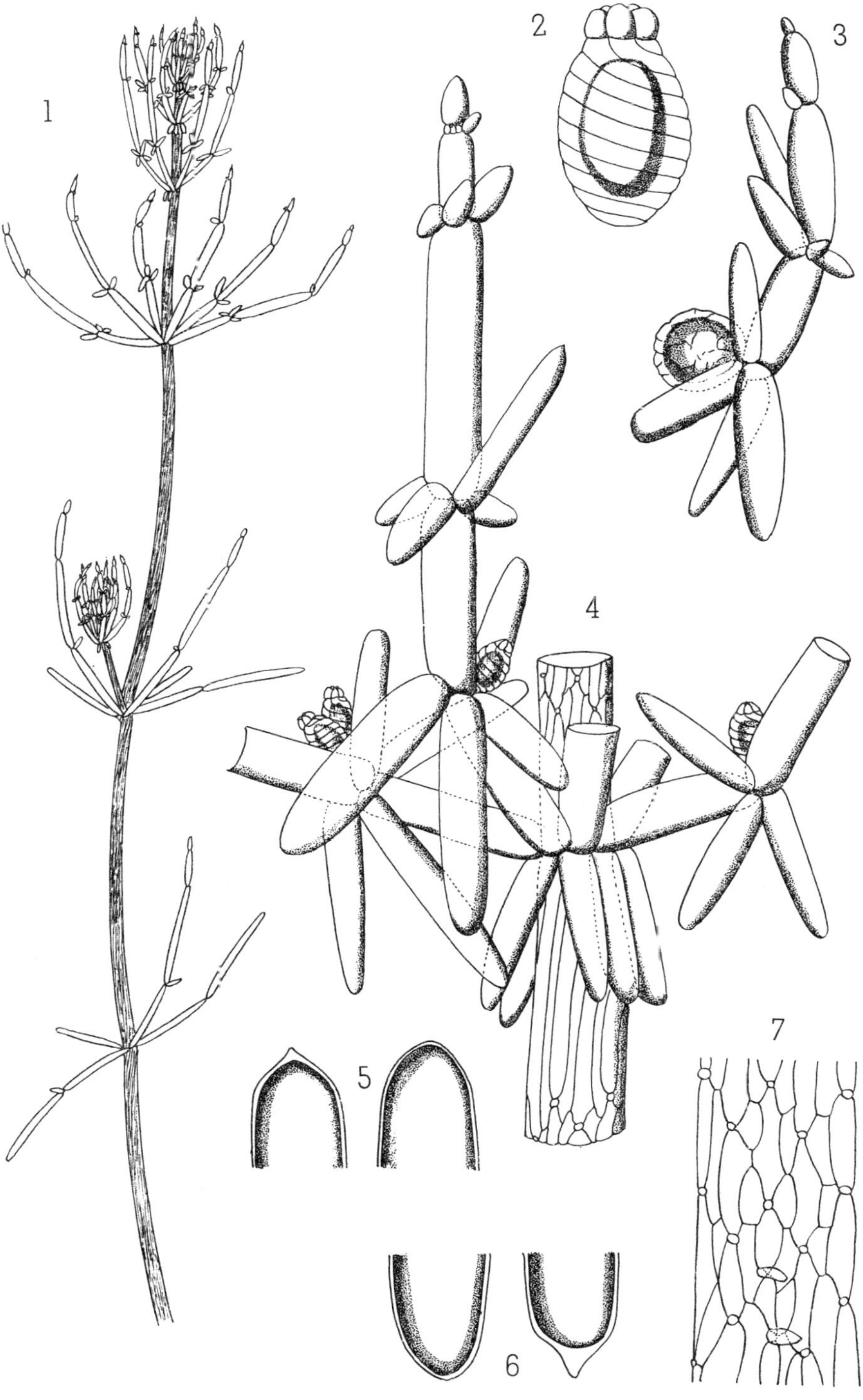
1
2
3
4
5
6
7

9. **Nitella flexilis** (L.) Ag.

(Nitella opaca, etc., see p. 61)

FIELD CLUES: common nitella with 1-forked branchlets; dactyls 1-celled with apices acute, apiculate or obtuse, but not long tapering; medium large (to 18"). Easily confused with N. acuminata.

DIAGNOSTIC FEATURES:
monoecious or dioecious
branchlets 1-forked, rarely 2-forked
dactyls 1-celled, apices acute, apiculate, or obtuse
heads absent, or lax; without mucus

HABITAT: forming loose "forests" or isolated clumps in deep, occasionally shallow, water of lakes, ponds, and rivers with mildly acid, and occasionally slightly brackish or alkaline, water.

VARIATIONS: considerable ecological and genetic variation, e.g., deep-water and over-wintering plants often blackish. Some unusual traits which occur are
- diffuse to compact stature; occasionally dendroid
- terminal whorls compacted into loose heads
- tiny size (to 1.5")
- monodactylous
- dioecious strains fairly common
- shortened branchlets
- irregular branchlets, occasionally 2-furcate

CLASSIFICATION: var. FLEXILIS, common, highly variable plants, with branchlets 1-furcate.
var. SPANIOCLEMA (J. Gr. & B.-W. ex B.-W.) R. D. W., the American specimens are small with irregular branchlets.

NOTE: dioecious strains common and often distinctive, occur sporadically throughout range.
NOTE: dactyl apex (fig. 5) is too blunt. Tips are usually much sharper.

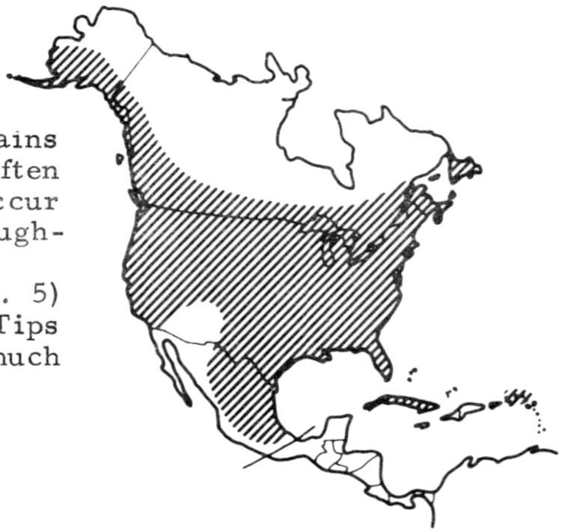

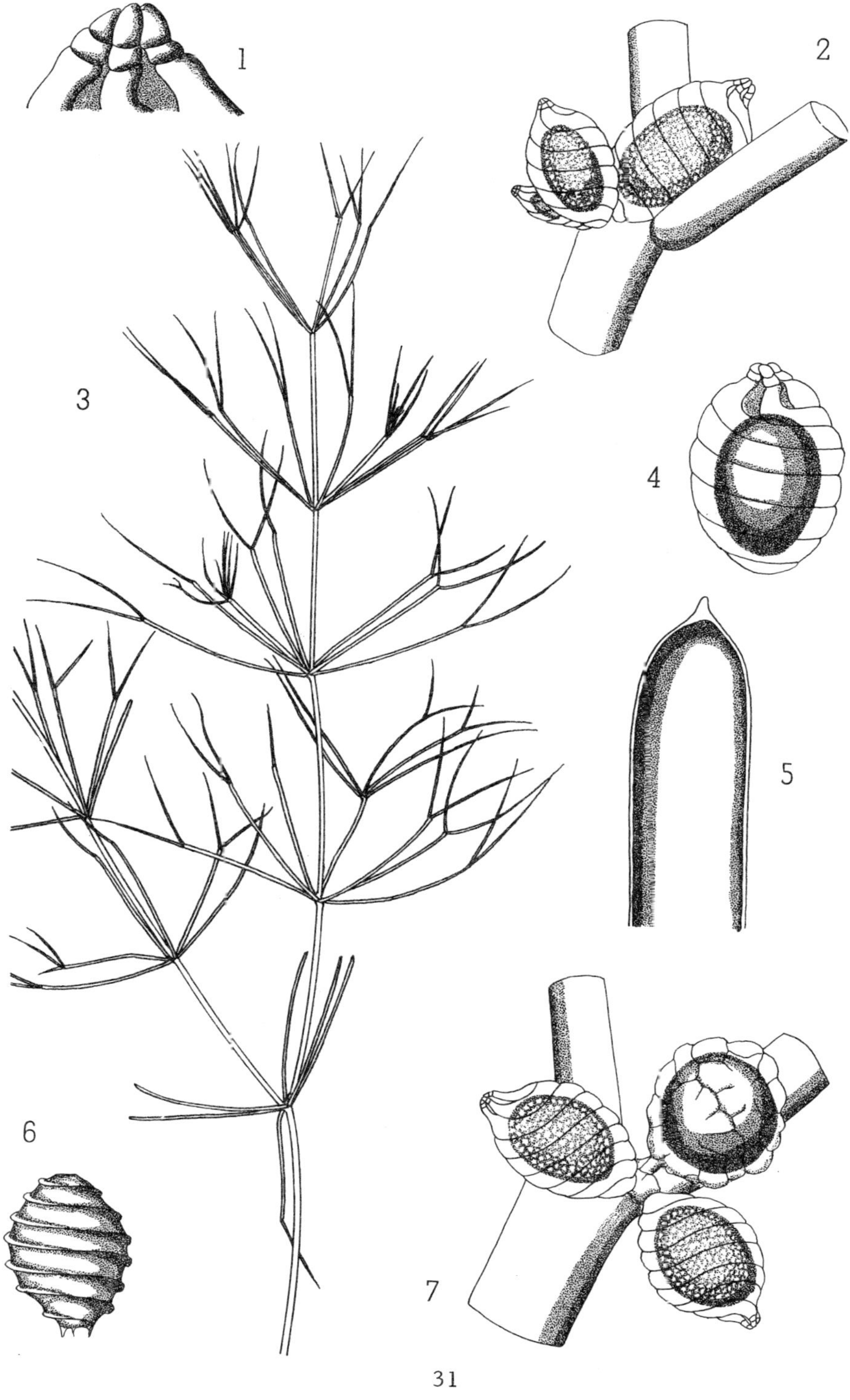
1
2
3
4
5
6
7

10. **Nitella macounii** (T. F. A.) T. F. A.

(Tolypella macounii)

FIELD CLUES: rare nitella, compact, with treelike stature, dark green color, 2-furcate branchlets, dactyls 1—2-celled; end-cell similar to basal cell in size, acute-tipped; small (12").

DIAGNOSTIC FEATURES:
- monoecious
- branchlets variable, (1—)2(—3)-furcate
- dactyls usually 1-celled, rarely 2—3-celled; end-cell of 2-celled branchlets not reduced, similar to a 1-celled dactyl
- heads absent, all whorls being compact
- gametangia very rarely at base of whorl

HABITAT: apparently isolated clumps in rivers of limestone areas.

VARIATIONS: slight.

CLASSIFICATION: nearly uniform, perhaps a single genotype (microspecies).

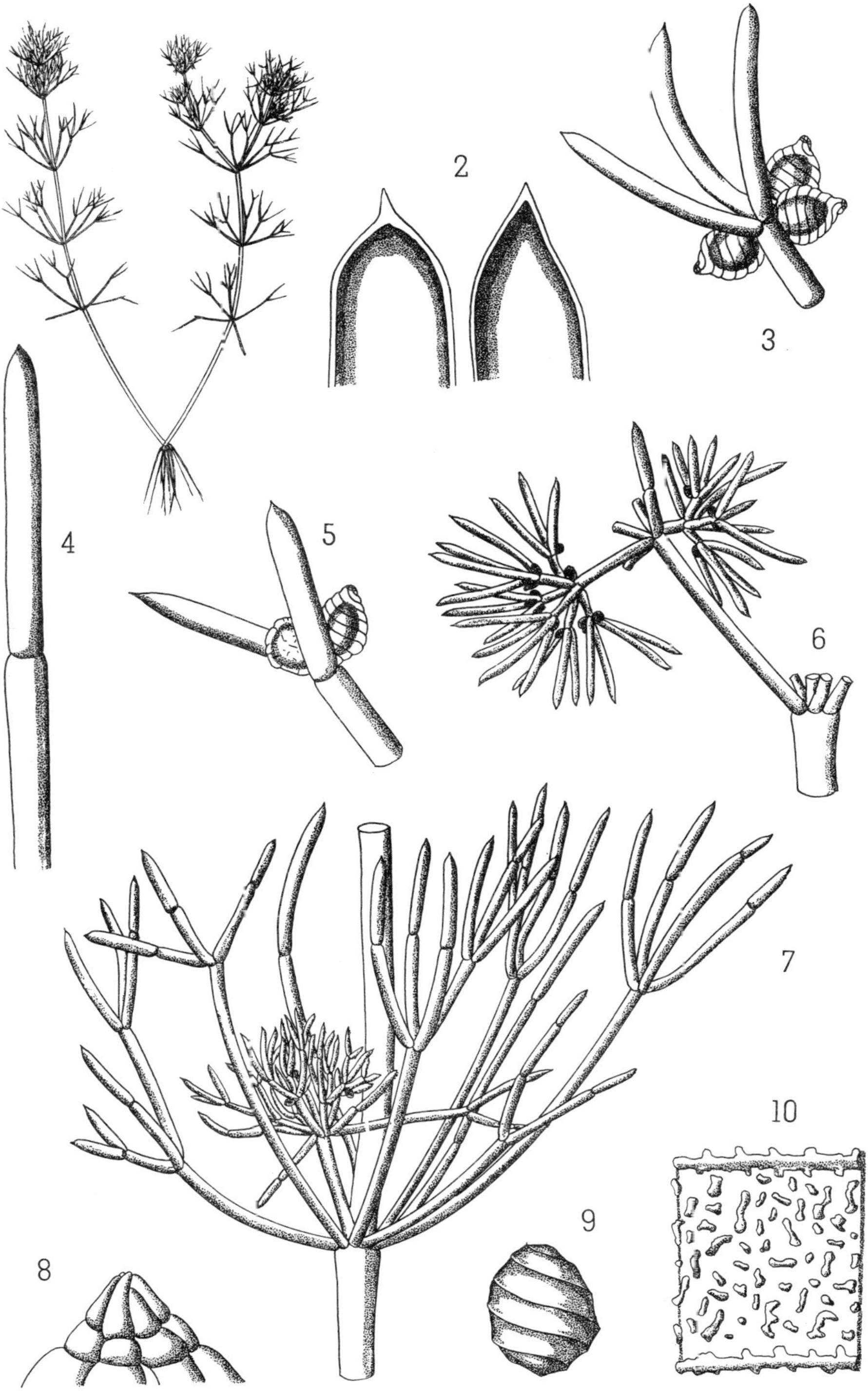
2
3
4
5
6
7
8
9
10

11. **Nitella acuminata** A. Br. ex Wallm.

(Nitella glomerulifera, N. subglomerata, etc., see p. 61)

FIELD CLUES: fairly common nitella with 1-forked branchlets, with dactyls 1-celled and their apices long tapering; medium large (to 18"). Easily confused with N. flexilis.

DIAGNOSTIC FEATURES:
- monoecious or dioecious
- branchlets 1-forked
- dactyls 1-celled, apices long tapering (acuminate)
- heads absent, weakly formed, or small and densely compact
- dactyls reduced

NOTE: differs from N. flexilis by having smaller mature oospores (225—340 μ vs. 375—575 μ long) with more striae (6—9 vs. 4—7) and narrower fossae (42—73 μ vs. 75—120 μ across).

HABITAT: fairly common in mildly acid or alkaline rivers; less common in standing water.

VARIATIONS: considerable ecological and great genetic variation. Apparently not blackened as occurs in N. flexilis. Some unusual traits which occur are
- lax, moderately compact heads, or tightly compact heads
- heads axial or terminal
- dendroid stature
- stout construction

CLASSIFICATION: var. ACUMINATA, the common form with lax or subcapitate structure.

var. GREENII R.D.W., with small, dense, axillary heads.

var. GOLLMERIANA (A. Br.) Zanev. ex R. D. W., branchlets look simple, but terminated with tiny dactyls barely visible to naked eye.

NOTE: dioecious strains numerous and distinctive, especially in Missouri and California.

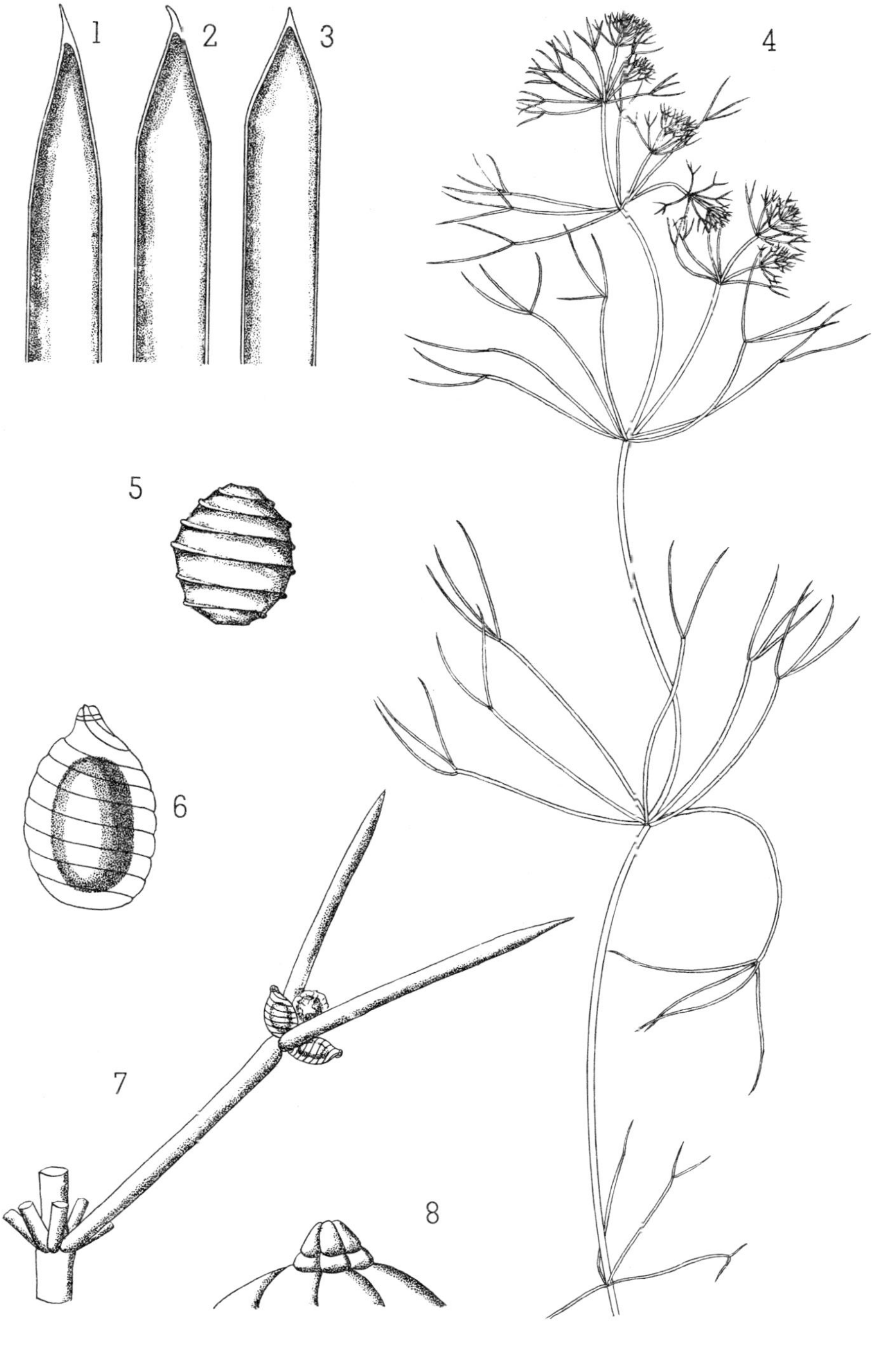
1
2
3
4
5
6
7
8

12. **Nitella bastinii** T. F. A.

(Nitella bastini)

FIELD CLUES: rare nitella with rigid stature, 1-furcate branchlets, 1-celled dactyls, and gray lime incrustation; medium size (to 12").

DIAGNOSTIC FEATURES:
- dioecious
- gametangia both at nodes of and at the base of branchlets
- branchlets 1-forked
- dactyls 1-celled, apices variable, blunt to acuminate
- heads compact, with mucus
- oospore membrane reticulate

HABITAT: known only from type locality, in shallow water of lake and pool.

VARIATIONS: slight.

CLASSIFICATIONS: nearly uniform, perhaps a single genotype (microspecies).

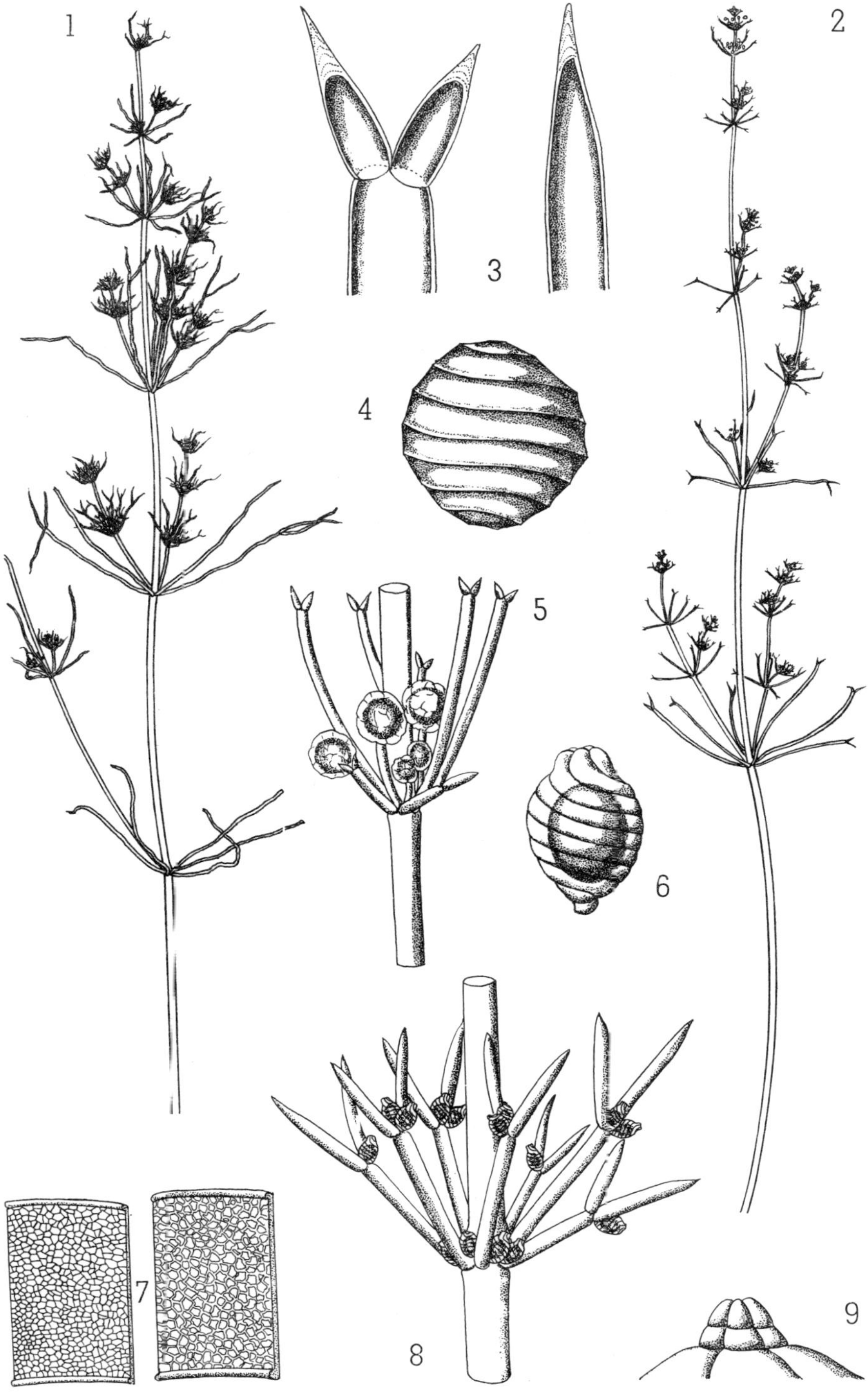
1
2
3
4
5
6
7
8
9

13. **Nitella clavata** Kütz.

(Nitella dilatata)

FIELD CLUES: common nitella usually somewhat succulent or of crisp texture; with swollen parts, bright green color, with compact, isolated whorls of short swollen branchlets; moderately large (to 18").

DIAGNOSTIC FEATURES:
monoecious
branchlets 1-forked, but heteroclemous with 1-celled accessory branchlets alternating with furcate branchlets
dactyls 1-celled, apices acuminate
heads generally absent, but branchlet whorls compact and headlike

HABITAT: common in shallow to fairly deep water of fresh-water lakes and ponds.

VARIATIONS: considerable ecological and genetic variation. Some unusual traits which occur are
small size
accessory branchlets few
branchlets shortened to nearly spheroid
fertile branches modified into spikelike head
internodes with banded incrustation

CLASSIFICATION: var. CLAVATA, the common robust form.
var. DILATATA (T. F. A.) R. D. W., small and slender, with few accessory branchlets.
NOTE: var. dilatata is easily confused with N. acuminata, but has larger oospores (300–480 μ long) and narrower fossae (59-65 μ) than N. acuminata (225—340 μ long, fossa 75—120 μ wide).

1
2
3
4
5
6
7
8

14. **Nitella cernua** A. Br.

(Nitella praelonga)

FIELD CLUES: uncommon robust nitella with tiny dactyls and tiny fertile heads with very dense mucus; large (to 6').

DIAGNOSTIC FEATURES:
- monoecious or dioecious
- branchlets appear simple, but have tiny terminal dactyls
- dactyls tiny, 1-celled
- heads small, .25—.75" long; with dense mucus
- gametangia restricted to fertile heads

HABITAT: large clumps in fairly shallow water of ponds, streams, and volcanic lakes.

VARIATIONS: fairly uniform.

CLASSIFICATION: var. CERNUA, the more southerly form, dioecious.
var. PRAELONGA (A. Br.) R.D.W., the more northerly form, monoecious, perhaps stouter (to 2800 μ) than var. cernua (axis to 2000 μ in diam.).

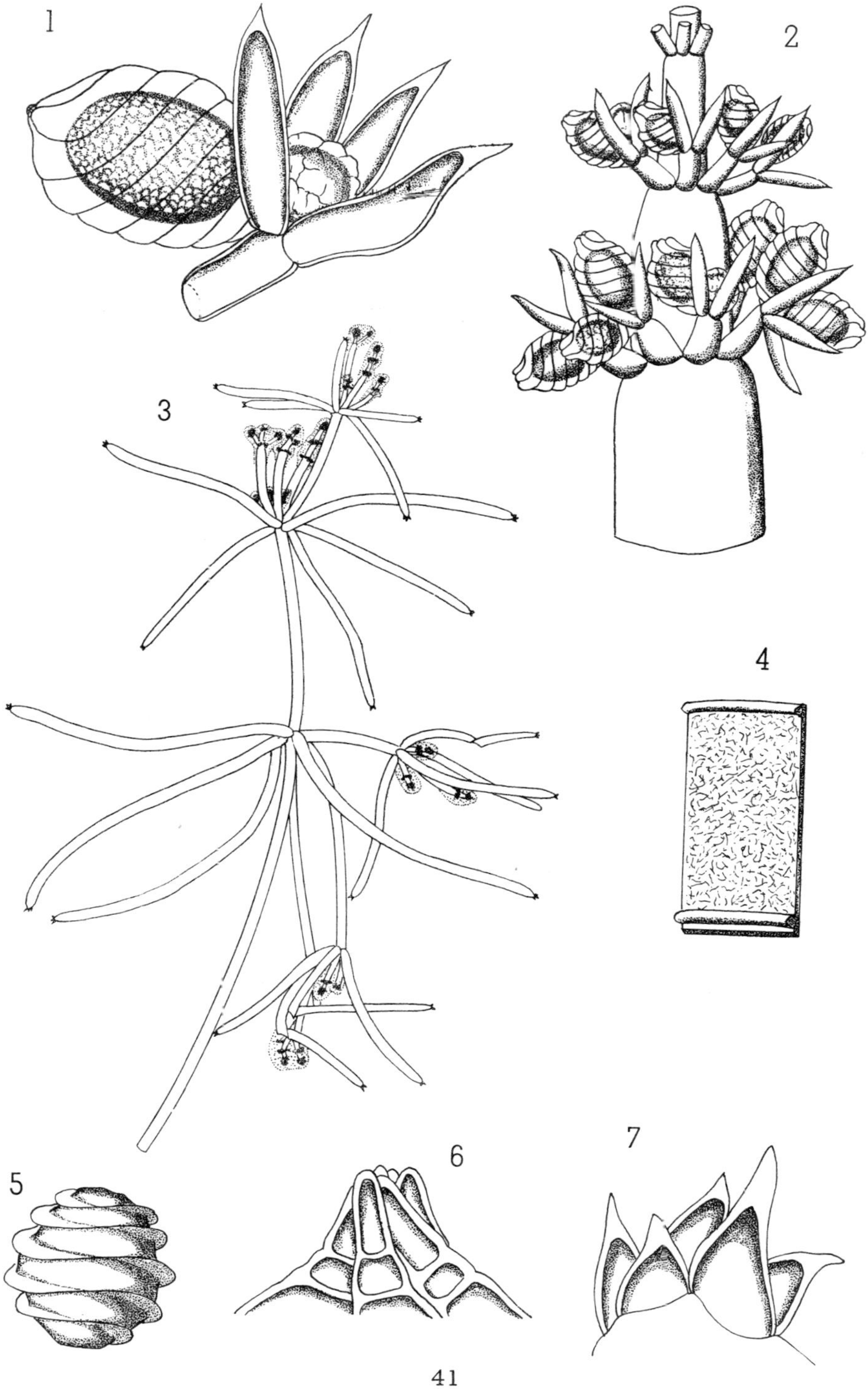
1
2
3
4
5
6
7

15. **Nitella capillata** A. Br.

FIELD CLUES: delicately divided nitella with distinctive terminal conelike heads with mucus; moderately small (to 12").

DIAGNOSTIC FEATURES:

monoecious

branchlets 2—4-furcate

dactyls 2—3-celled; end-cell variable, similar to basal cells and blunt, or short and pointed

heads distinct, .25-.75" long; with mucus

HABITAT: in loose "forests" in shallow water of ponds and streams, soft-water areas.

VARIATIONS: fairly uniform.

CLASSIFICATION: not divided into varieties; perhaps a single genotype (microspecies).

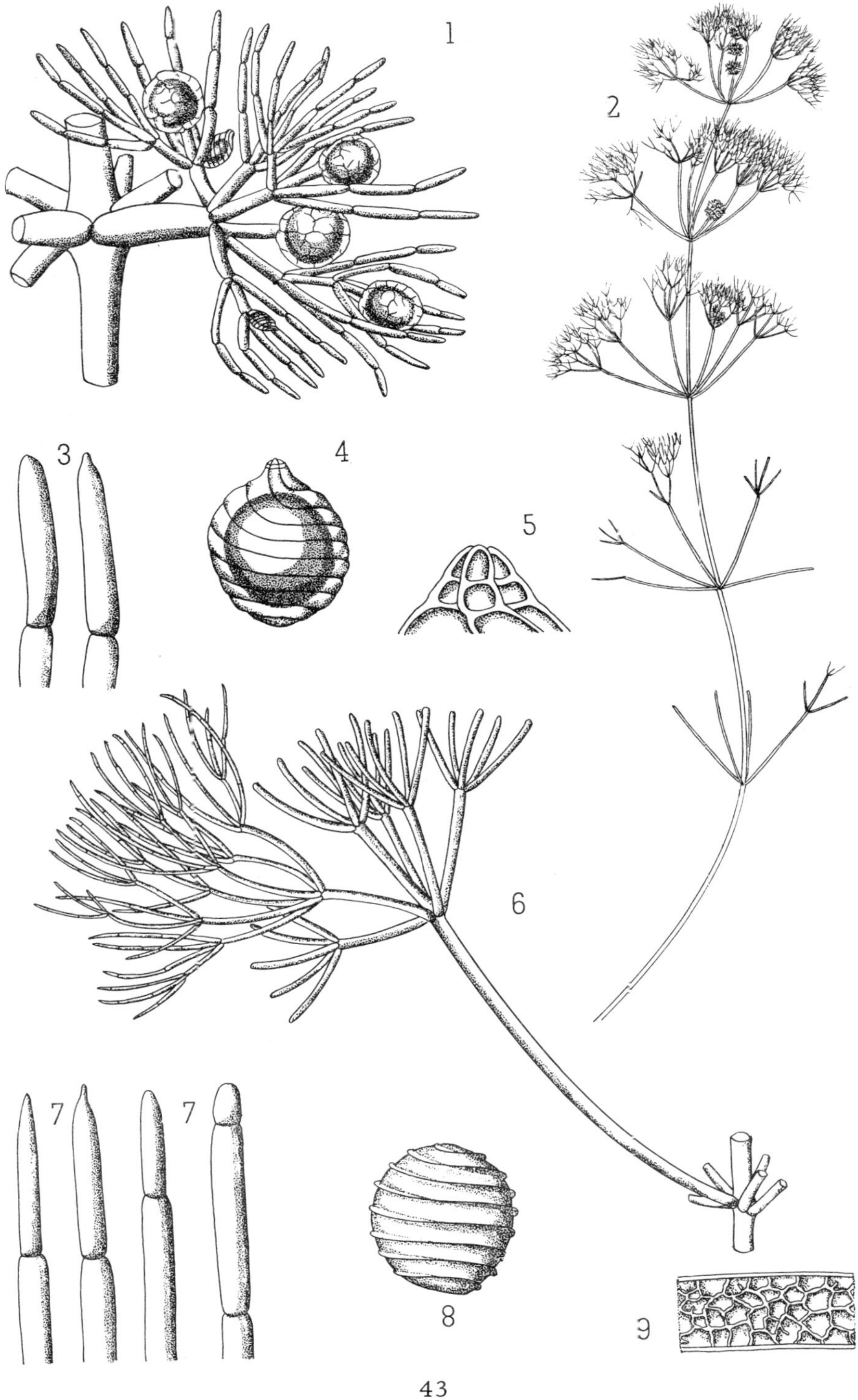
1
2
3
4
5
6
7
7
8
9

16. Nitella furcata (Roxb. ex Bruz.) Ag.

(Nitella megacarpa, N. mucronata, etc., see p. 62)

FIELD CLUES: uncommon, but locally abundant, robust nitella with broad structure and finely divided branchlets; moderately large (to 18").

DIAGNOSTIC FEATURES:
- monoecious
- branchlets 2—4-furcate
- dactyls 2(—3)-celled, with tiny, mucronate end-cell; entire set of dactyls commonly tiny
- heads absent or obscure, but young whorls may resemble heads
- oospore membrane coarsely reticulate or papillate
- gametangia solitary to aggregate (spores often mature very late)

HABITAT: moderately shallow, in soft-water ponds and streams with rich mud bottom.

VARIATIONS: considerable ecological and genetic variation. Some unusual traits which occur are
- oogonia solitary, geminate, or aggregate
- dactyls all normal-sized
- oospore membrane reticulate or papillate
- sterile branchlets only 1—2-furcate
- oospores small, medium, or large
- percurrent central secondary ray
- dactyls 3-celled

(continued on page 46)

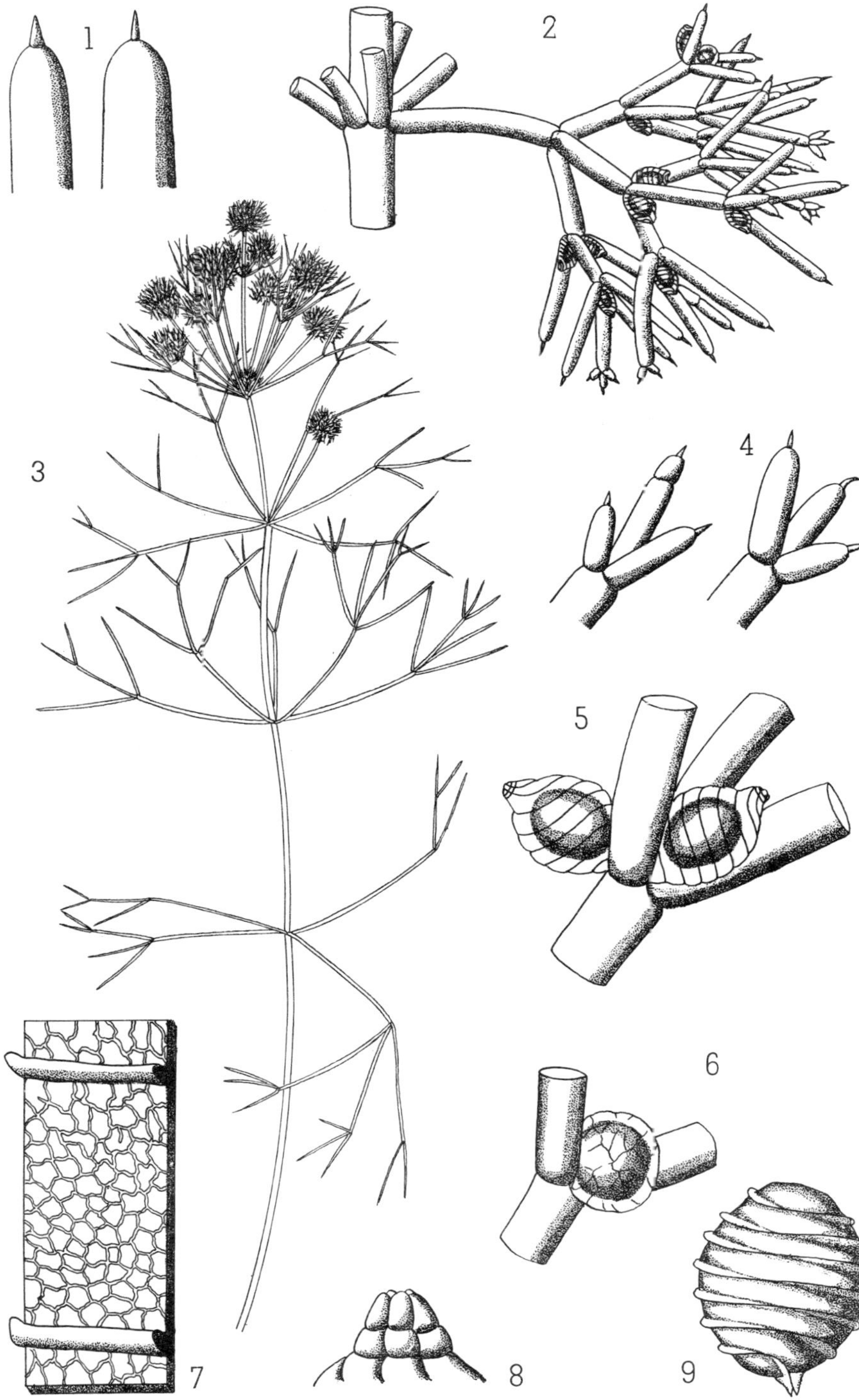
1
2
3
4
5
6
7
8
9

CLASSIFICATION: subsp. FURCATA, uncommon form with most dactyls small; includes

- var. FURCATA, with most dactyls small;
- var. SIEBERI (A. Br.) R. D. W., has some dactyls small.

subsp. MUCRONATA (A. Br.) R. D. W., common form with dactyls normal, predominantly 2-celled; includes

- var. MUCRONATA, common, branchlets 2—3-furcate;
- var. CAPITELLATA (A. Br.) R.D. W., questionable and poorly known taxon, possible N. gracilis, with branchlets 1—2-furcate; dactyls commonly 3-celled and elongate.

subsp. MEGACARPA (T. F. A.) R. D. W., common, robust, with occasional small dactyls, aggregate oogonia, and oospore membrane papillate.

Plate on page 47

Nitella furcata ssp. megacarpa (T. F. A.) R. D. W.

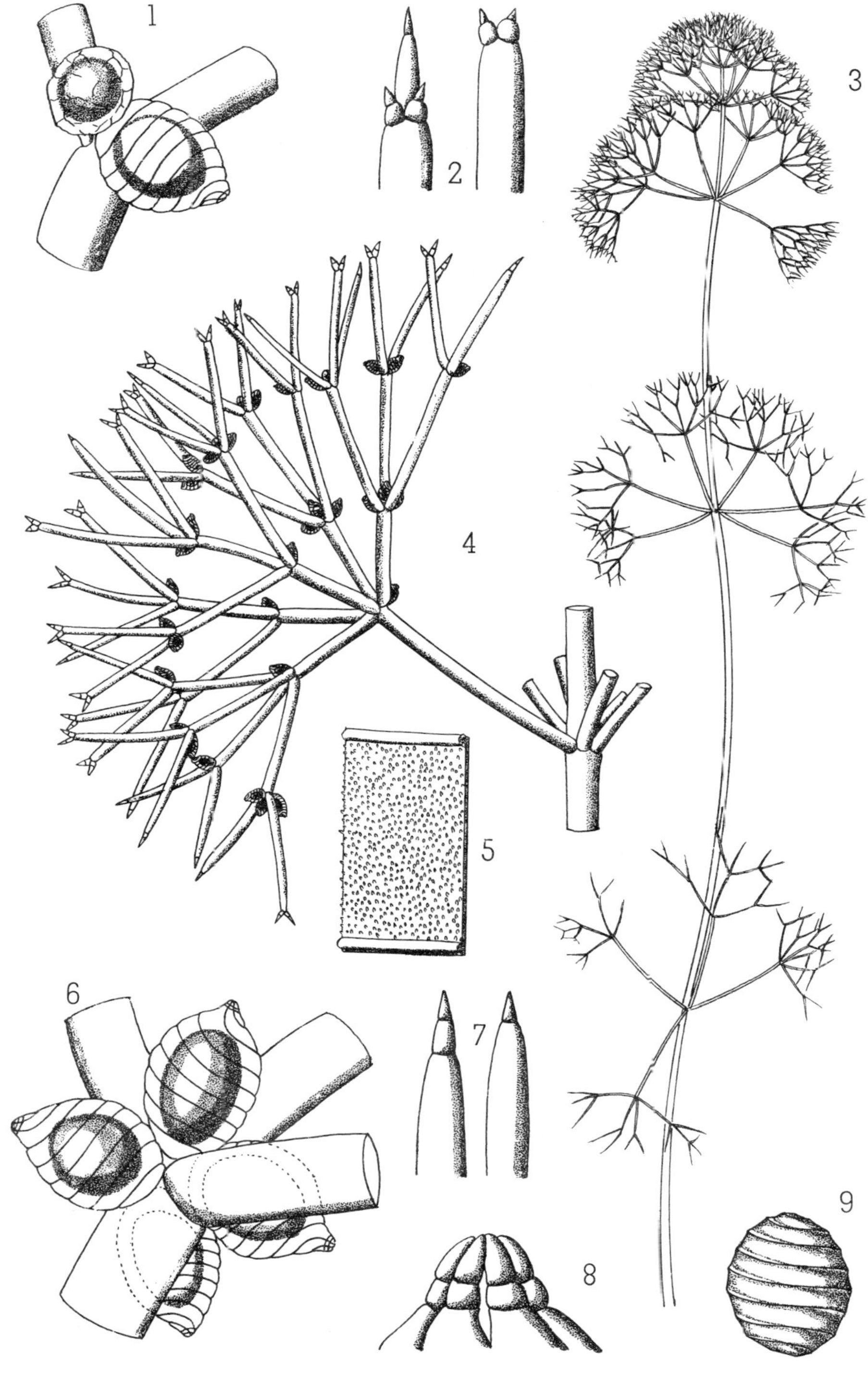
1
2
3
4
5
6
7
8
9

17. **Nitella tenuissima** (Desv.) Kütz.

(Nitella transilis)

FIELD CLUES: locally common, small delicate nitella with much-divided branchlets, but without heads or mucus; generally small (to 10").

DIAGNOSTIC FEATURES:
- monoecious
- branchlets 2—4-furcate, with central secondary ray
- dactyls 2-celled; end-cell conical, acute
- heads absent though sterile whorls may all be compact and headlike
- oospore membrane coarsely reticulate (spores often mature late)
- gametangia rare at lowest branchlet node

HABITAT: common in fairly shallow water of soft-water lakes and ponds.

VARIATIONS: considerable ecological and some genetic variation. Some unusual traits which occur are
- whorls compacted and isolated (moniliform)
- branchlet whorls lax and fanlike
- large examples

CLASSIFICATION: American material is not divided into varieties.

NOTE: easily confused with N. gracilis, but N. tenuissima has dactyls generally only 2-celled, a central secondary branchlet ray, and reticulate oospore membrane.

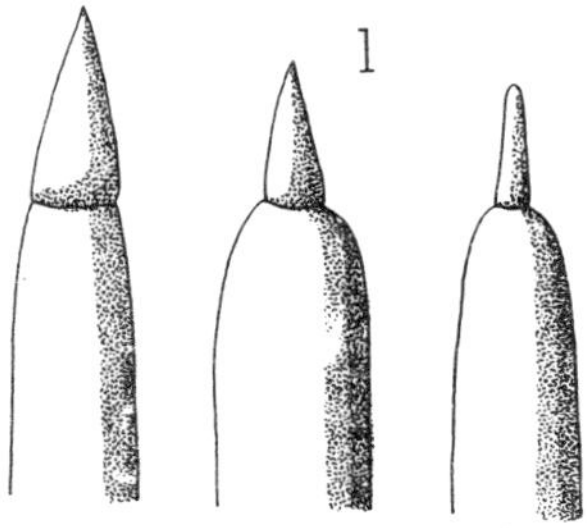

1

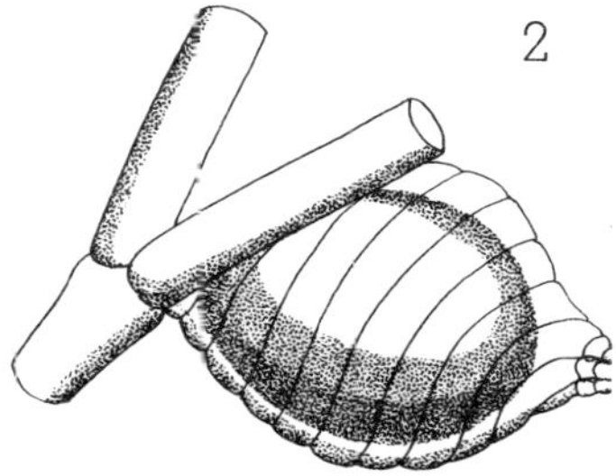

2

3

4

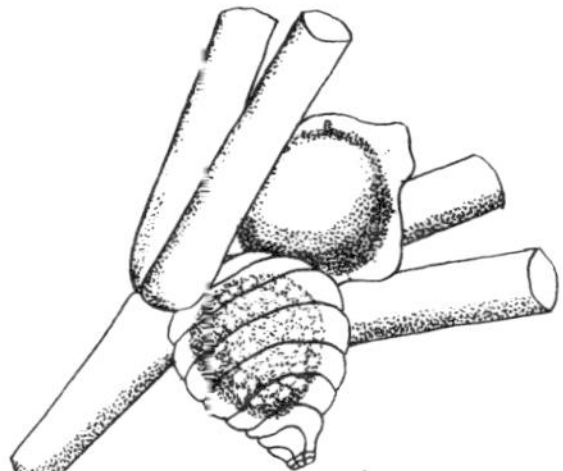

5

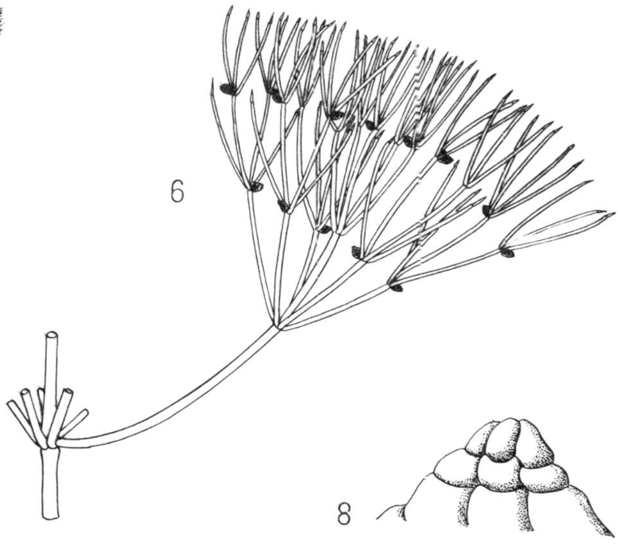

6

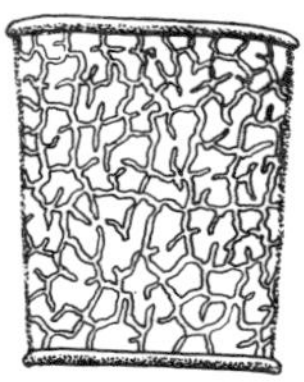

7

8

18. **Nitella gracilis** (Sm.) Ag.

(Nitella batrachosperma, N. intermedia, N. morongii, etc., see p. 62)

FIELD CLUES: locally common, small delicate nitella with much-divided branchlets and more or less distinct heads with a suggestion of mucus; generally small (to 10").

DIAGNOSTIC FEATURES:
- monoecious
- branchlets 2—5-furcate, rarely with central secondary ray
- dactyls 2- or 2—3-celled; end-cell conical, acute
- heads generally more or less distinct, commonly with mucus
- oospore membrane granulate, punctate, or pitted
- gametangia common at lowest branchlet node

HABITAT: common in fairly shallow water of lakes and rivers.

VARIATIONS: considerable ecological and much genetic variation. Some unusual traits which occur are
- dactyls all 2-celled
- with central secondary branchlet ray
- heads distinct, with definite mucus
- oospore membrane vermiferous or pitted
- dactyl end-cell nearly mucronate
- dactyls somewhat abbreviated
- nodes swollen

CLASSIFICATION: var. GRACILIS, common form without special features.
- var. LEIBERGII (T.F.A.) R. D. W., 2-celled dactyls and compact head without mucus.
- var. CONFERVACEA Bréb., 2-celled dactyls, tiny plants (to 1.5" high), with suggestion of mucus.
- var. ASAGRAYANA (Schaffn. ex Nordst. in A. Br. & Nordst.) R. D. W., dactyls 2-celled, with distinct mucus.

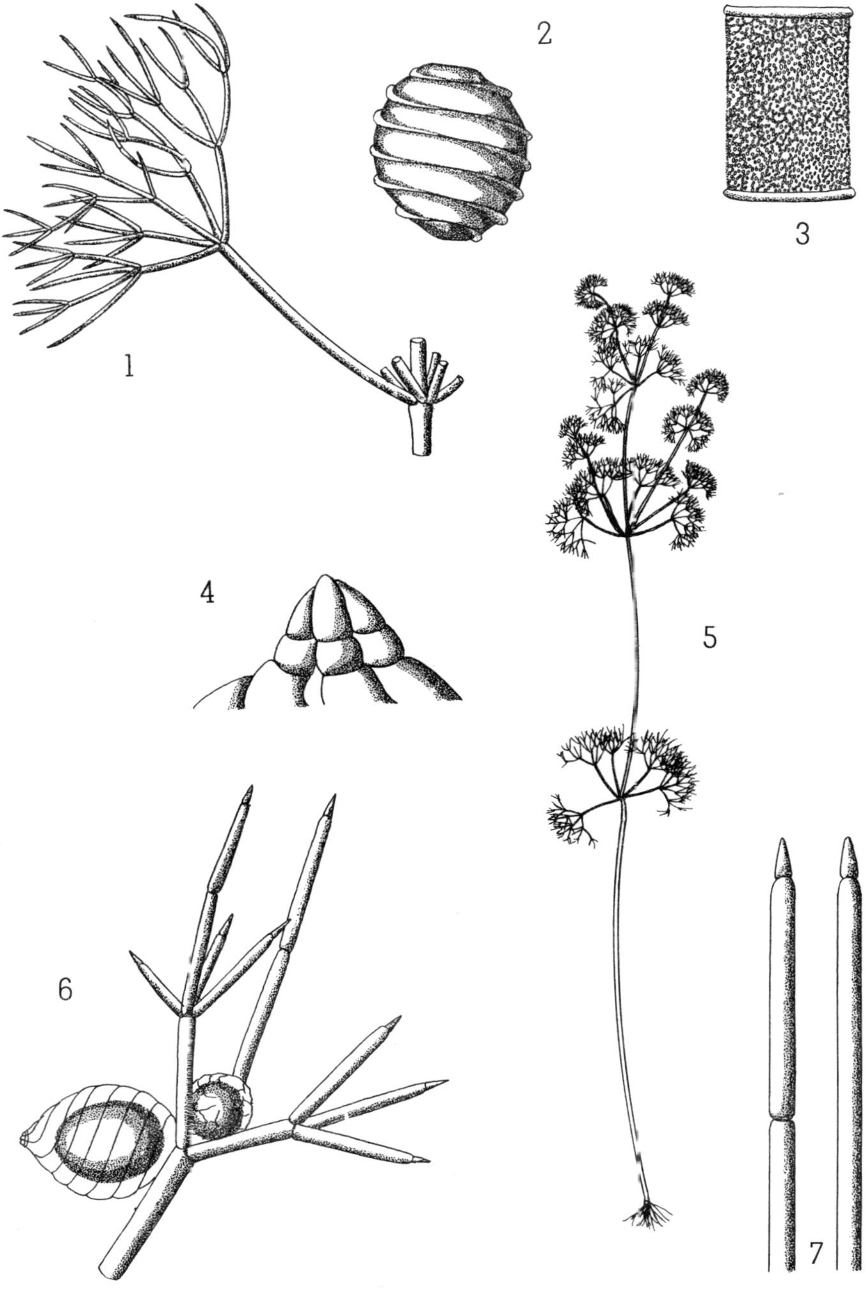
1
2
3
4
5
6
7

19. **Nitella hyalina** (DC.) Ag.

FIELD CLUES: locally common, small delicate nitella with much-divided branchlets, more or less distinct heads with mucus, dark green color, heteroclemous branchlets; commonly incrusted; generally small (to 10").

DIAGNOSTIC FEATURES:
monoecious
branchlets 2—3-furcate, heteroclemous with an inner whorl of accessory branchlets at each whorl
dactyls 2-celled; end-cells somewhat mucronate
heads more or less distinct, with mucus
oospore membrane granulate

HABITAT: sparse, locally common, in isolated clumps in ponds and lakes, apparently mostly in hard or very weakly brackish water.

VARIATION: fairly uniform.

CLASSIFICATION: North American examples not divided into varieties, although var. MAXIMA (A. Br. ex Mig.) R. D. W., a large form (1.5—2'), presumably European, possibly may appear in Louisiana.

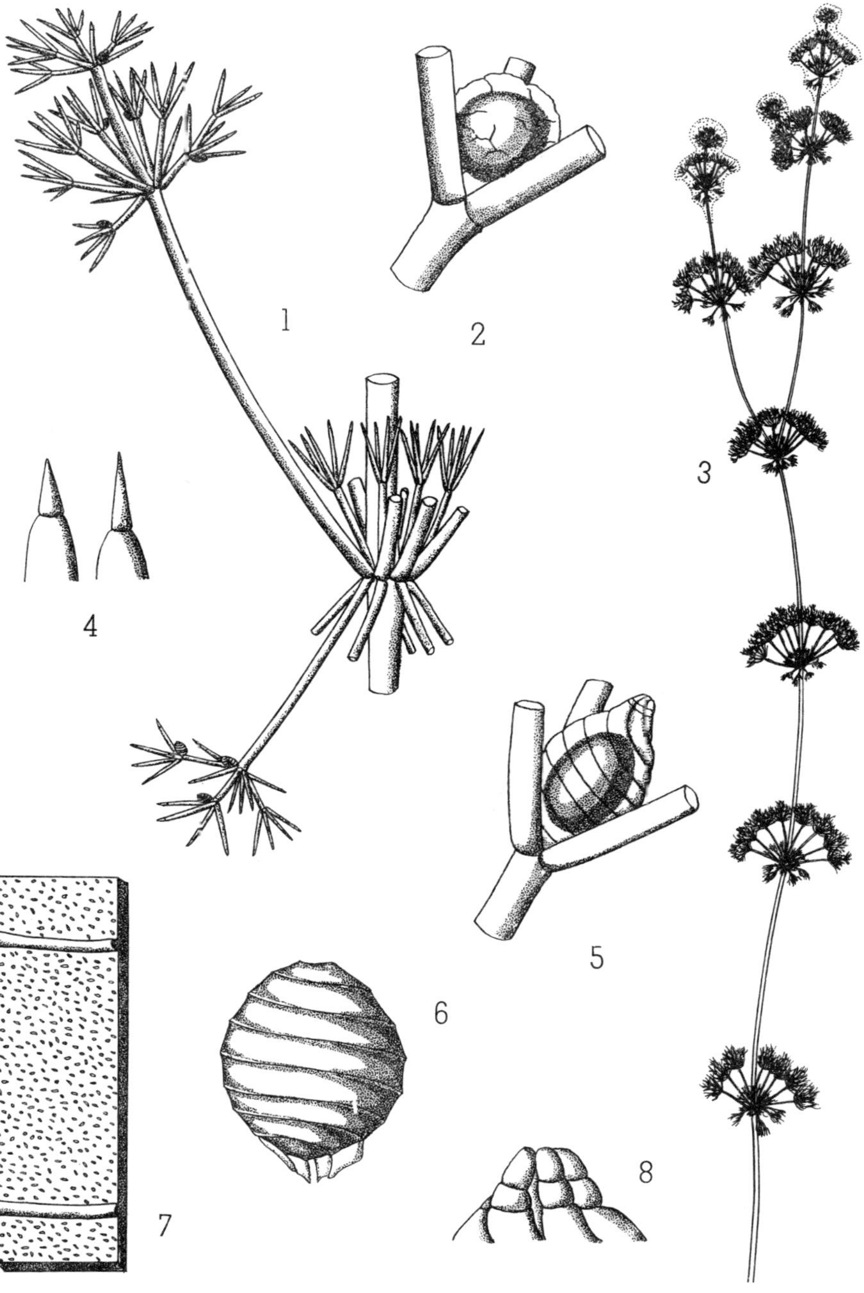
1
2
3
4
5
6
7
8

20. **Nitella translucens** (Pers.) Ag.

(Nitella axillaris)

FIELD CLUES: uncommon, robust nitella with branchlets appearing simple but terminated with tiny dactyls, and with compact axillary heads without mucus; large (to 2. 5').

DIAGNOSTIC FEATURES:

monoecious
branchlets 1-furcate, but appearing simple by having tiny terminal dactyls; very rarely 2-furcate
dactyls of sterile branchlets minute, 2-celled
heads small and compact, axillary, without mucus
oospore membrane coarsely reticulate

HABITAT: isolated clumps in shallow water of ponds and lakes, generally in hard or slightly brackish water.

VARIATIONS: considerable ecological and some genetic variation. Some unusual traits which occur are

sterile branchlets 2-furcate
dactyls end-cells bifid (with lateral spurs)
gametangia occurring on sterile branchlets
dactyls 3-celled
oospore membrane finely instead of coarsely reticulate

CLASSIFICATION: the American examples conform to var. AXILLARIS (A. Br.) R. D. W. as described above.

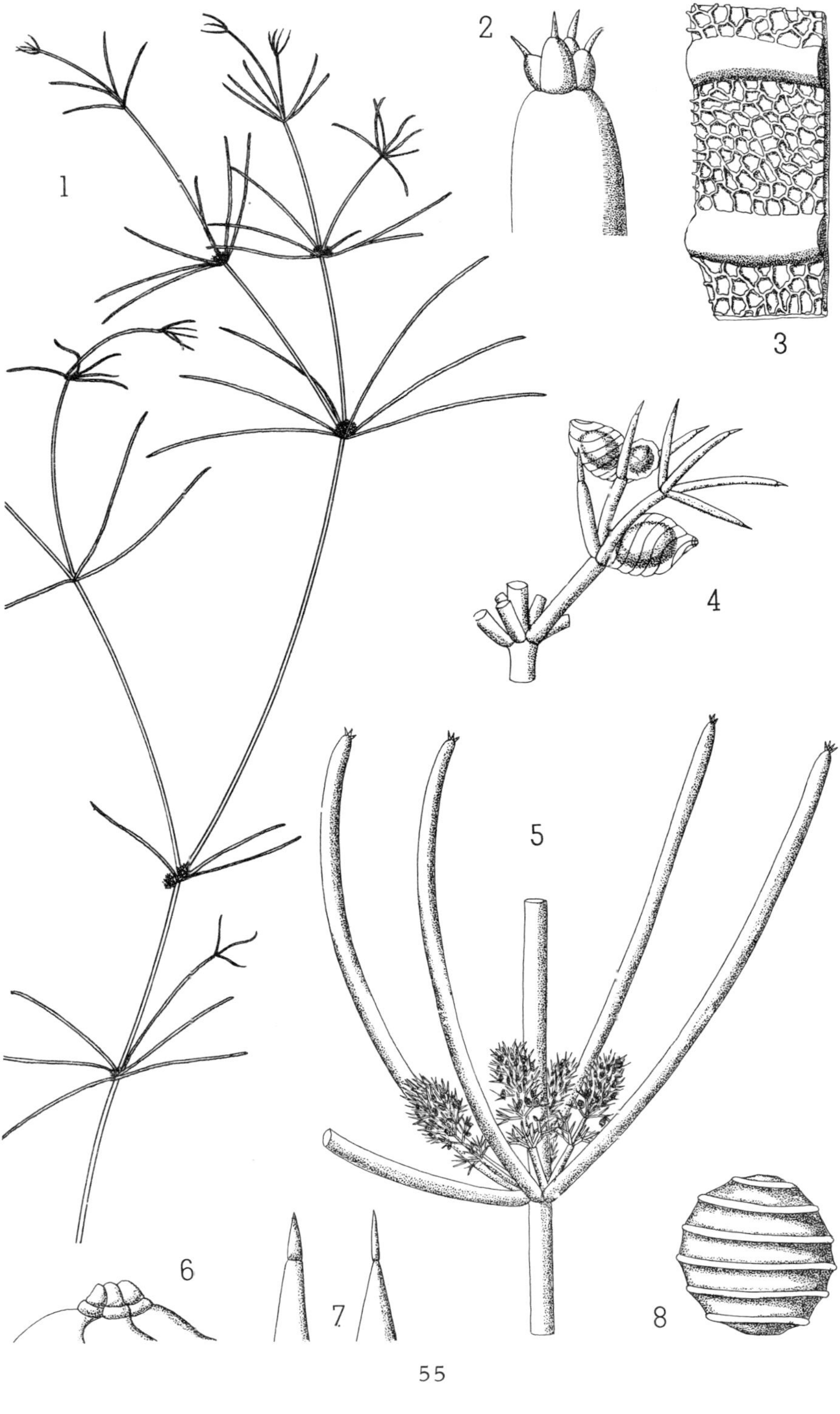
1
2
3
4
5
6
7
8

21. Tolypella nidifica (O. Müll.) A. Br.

(Tolypella comosa, T. glomerata, T. hispanica, etc., see p. 62)

FIELD CLUES: fairly common, with gray lime incrustation, coarse bird's-nestlike heads, and blunt branchlet end-cells; medium-sized (to 15").

DIAGNOSTIC FEATURES:

monoecious, rarely dioecious
branchlets monopodial, not furcate, but with 1(—2) ray-bearing nodes
end-cell rounded, blunt
lateral branchlet rays 2—6-celled
heads few to numerous

HABITAT: isolated shoots in shallow to fairly deep water, hard-water ponds, lakes, and streams.

VARIATIONS: considerable ecological and some genetic variation. Some unusual traits which occur are

branchlets either attenuate or allantoid
branchlet end-cells elongated
tiny in size (1.25")
sterile branchlets 1—3-divided

CLASSIFICATION: var. NIDIFICA, common form with tapering branchlets and small end-cell.

var. PORTERI (Daily) R. D. W., rare, tiny, dioecious, with large antheridia.

var. GLOMERATA (Desv. in Lois.) R. D. W., fairly common, with branchlets not tapering, end-cell large, sausage-shaped.

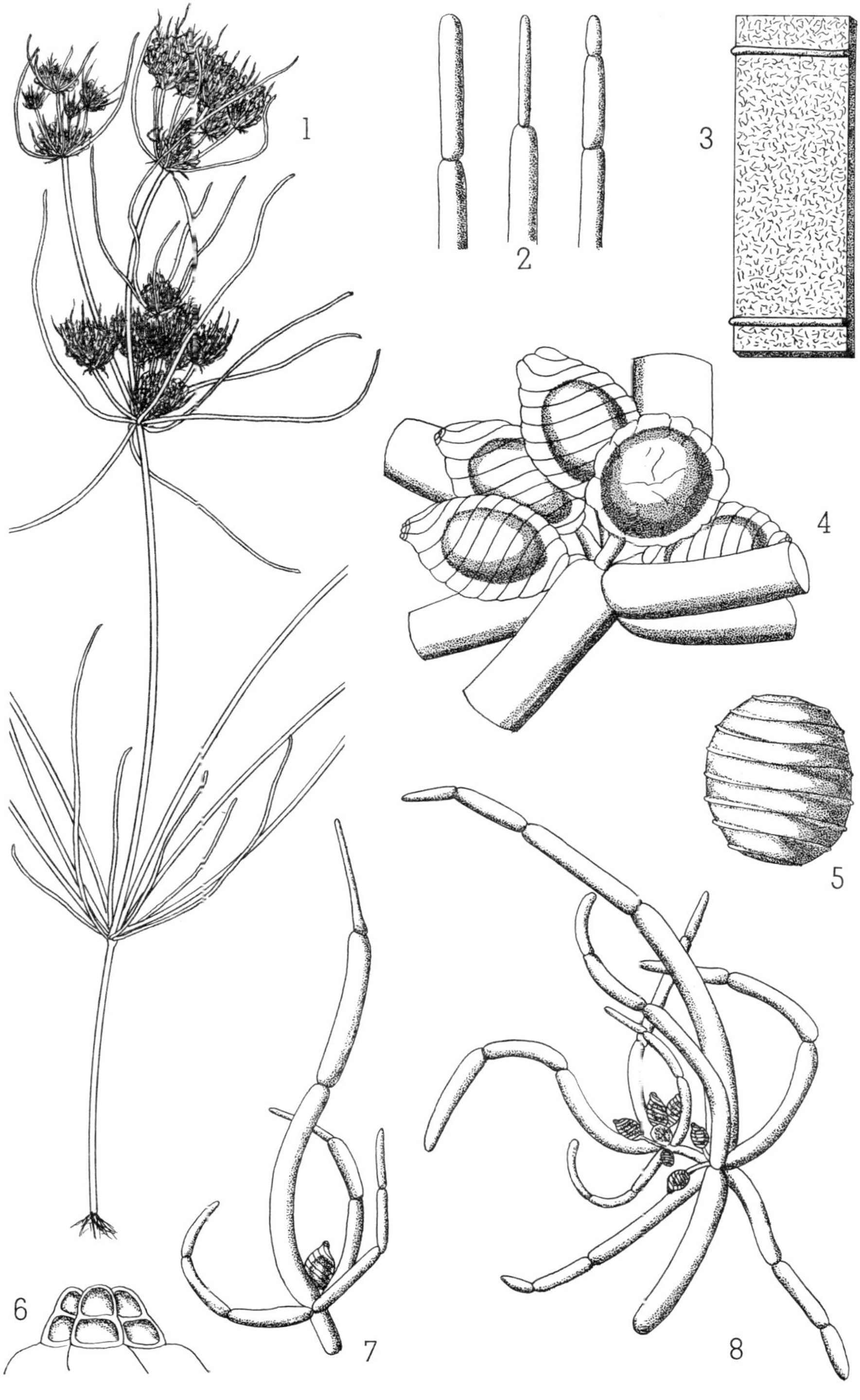
1
2
3
4
5
6
7
8

22. **Tolypella intricata** (Trent. ex Roth) Leonh.

(Tolypella intertexta, T. prolifera, etc., see p. 62)

FIELD CLUES: fairly common, generally with gray lime incrustation, coarse bird's-nestlike heads, and branchlet end-cells reduced and acute-tipped; medium large (to 18").

DIAGNOSTIC FEATURES:
- monoecious
- branchlets monopodial, not furcate, but with 1—more ray-bearing nodes; sterile branchlets commonly appearing divided
- end-cells reduced, acute-tipped
- lateral branchlet rays commonly with secondary rays
- heads few to numerous
- adventitious branchlets common

HABITAT: isolated shoots in shallow to fairly deep water, hard-water ponds, lakes, and streams.

VARIATIONS: considerable ecological and some genetic variations. Some unusual traits which occur are
- sterile branchlets divided 1—2 times
- stout structure
- with numerous tiny adventitious laterals
- with numerous sterile branchlets
- fertile whorls compacted into spike

CLASSIFICATION: American taxa not divided into varieties.

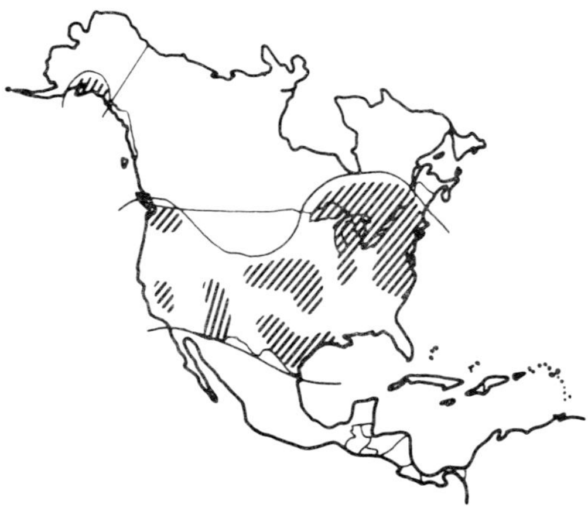

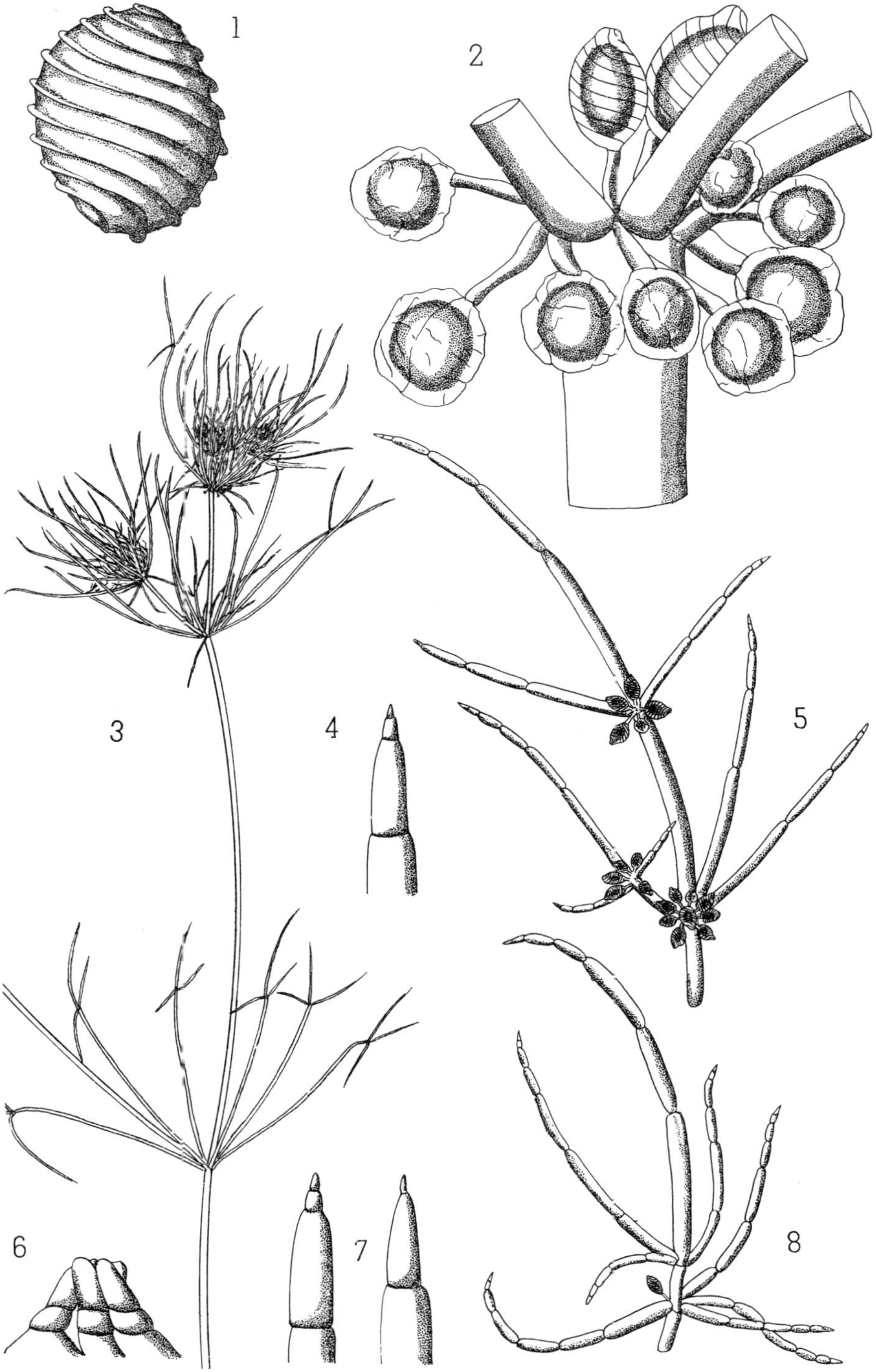
1
2
3
4
5
6
7
8

OTHER NAMES ASSOCIATED WITH EACH SPECIES IN AMERICAN LITERATURE

NOTES: Names not published as species names but only as microspecies (cf. Wood, 1965: 763 ff.) are so marked, e.g., under C. vulgaris see "(microsp.) C. hispidula."

Dioecious strains are marked "dioec."

1. (p. 12)
   CHARA VULGARIS
   C. contraria
   C. crassicaulis
   C. excelsa
   C. fetidissima
   C. foetida
   (microsp.) C. hispidula
   (microsp.) C. howeana
   C. hypnoides
   C. inconnexa
   (C. intermedia, misidentified)
   C. intumescens
   C. kieneri
   C. schaffneri
   C. tehuacanensis

2. (p. 14)
   CHARA CANESCENS (dioec.)
   C. crinita (dioec.)
   C. evoluta
   C. hirsuta

3. (p. 16)
   CHARA GLOBULARIS
   C. aspera (dioec.)
   C. delicatula
   C. fragilis
   C. hedwigii
   C. leptosperma
   C. macounii
   C. pulchella
   C. verrucosa
   C. virgata

4. (p. 18)
   CHARA ZEYLANICA
   C. brittonii
   C. carmenensis
   C. compacta
   C. conjungens
   C. cruegeriana
   C. cubensis
   C. depauperata
   (microsp.) C. diaphana
   C. domingensis
   C. elegans
   C. fertilissima
   C. filicaulis
   C. foliolosa
   C. formosa
   C. guadeloupensis
   C. guatemalensis
   C. gymnopus
   C. haitensis
   C. hicksiana
   C. hicksii
   C. humboldtii
   C. inconstans
   C. indica
   C. kenoyeri (dioec.)
   C. martiana Nordst. (dioec.)
   C. martiana Wallm.
   C. michauxii
   C. mühlenbergii
   C. oerstediana
   C. palmeri
   C. polyphylla
   C. rusbyana (dioec.) (S. America)
   C. sabulosa

C. sanctae-margaritae
C. sejuncta
(microsp.) C. tenuifolia
C. trichacantha

5. (p. 22)
CHARA BRAUNII
C. coronata
C. foliolosa Schweinitz
C. pallida
C. schweinitzii
Nitella coronata

6. (p. 24)
CHARA FIBROSA
C. coronatiformis
C. crinitiformis
C. curtissii
C. gymnopitys
C. hydropitys
C. keukensis
C. liebmannii
C. mexicana
C. morongii
C. myriophylla
C. robbinsii
C. schneckii
C. septentrionalis (ined.)

7. (p. 26)
CHARA HORNEMANNII (dioec.)
C. longifolia (dioec.)
C. nordhoffiae (dioec.)
(Nitellopsis bulbillifera?)
(Nitellopsis obtusa?)

8. (p. 28)
CHARA BUCKELLII (dioec.)

9. (p. 30)
NITELLA FLEXILIS
Acinaria coccifera
(microsp.) N. atkahensis
(microsp.) N. attenuata
N. californica (dioec.)
(N. capillaris, misidentified)
(N. capitata, misidentified)
(microsp.) N. dailyana
N. laxa
N. mexicana
N. monodactyla (dioec. ?)
N. montana (dioec.)
N. opaca (dioec.)
(microsp.) N. opacoides
N. spanioclema

10. (p. 32)
NITELLA MACOUNII
Tolypella macounii

11. (p. 34)
NITELLA ACUMINATA
Acinaria flexuosa
Chara glomerulifolia
N. belangeri
N. blankinshipii (dioec.)
N. glomerulifera
N. gollmeriana
(microsp.) N. greenii
N. lindheimeri
N. missouriensis (dioec.)
N. occidentalis (dioec.)
N. stellaris
N. subglomerata
N. subspicata

12. (p. 36)
NITELLA BASTINII (dioec.)
N. bastini
N. stricta (ined.)

13. (p. 38)
NITELLA CLAVATA
Chara clavata
N. dilatata

14. (p. 40)
NITELLA CERNUA (dioec.)
Acinaria salicifolia
N. gelatinosa Halst.
N. praelonga

15. (p. 42)
NITELLA CAPILLATA

16. (p. 44)
NITELLA FURCATA
Chara capitata Ell.
C. elliotii (ined.)
Leiacina capitata
N. capitellata
N. dictyosperma
N. flabellata
N. glaziovii
N. megacarpa
N. microcarpa
N. mucronata
N. oligospira
N. pygmaea

17. (p. 48)
NITELLA TENUISSIMA
N. transilis

18. (p. 50)
NITELLA GRACILIS
N. annularis
N. asagrayana
N. batrachosperma
N. confervacea
N. intermedia
N. leibergii
N. maxceana
N. morongii
N. nordstedtiana

19. (p. 52)
NITELLA HYALINA
(microsp.) N. engelmanni
N. formosa

20. (p. 54)
NITELLA TRANSLUCENS
N. axillaris
(microsp.) N. caroliniana

21. (p. 56)
TOLYPELLA NIDIFICA
Chara glomerata
Nitella glomerata
T. comosa
T. glomerata
T. hispanica (dioec.)
T. longicoma
(microsp.) T. porteri (dioec.)
(microsp.) T. subdivisa

22. (p. 58)
TOLYPELLA INTRICATA
Nitella polysperma
T. californica
T. fimbriata
T. intertexta
T. prolifera
(microsp.) T. spicata
T. stipitata

KEY

1a. With cortication 2
1b. Without cortication 11

2a. Branchlets corticated, at least in part ... 3
2b. Branchlets naked 7

3a. Basal branchlet segment corticated 4
3b. Basal branchlet segment naked 6

4a. Spine-cells fascicled; axial cortex haplostichous. . .CHARA CANESCENS, p. 14
4b. Spine-cells solitary or rarely geminate; cortex 2—3-corticated 5

5a. Axial cortex 2-corticated. CHARA VULGARIS, p. 12
5b. Axial cortex 3-corticated .... CHARA GLOBULARIS, p. 16

6a. Stipulodes in 1 tier .... CHARA FIBROSA var. HYDROPITYS, p. 24
6b. Stipulodes in 2 tiers .... CHARA ZEYLANICA, p. 18

7a. Stipulodes in 2 tiers 8
7b. Stipulodes in 1 tier 9

8a. Axial cortex 2-corticated .... CHARA VULGARIS (gymnophyllous forms), p. 12
8b. Axial cortex 3-corticated.............. CHARA ZEYLANICA var. DIAPHANA, p. 18

9a. Small plants, to 6 in. high; axes to 400 μ in diameter; monoecious . . . CHARA FIBROSA, p. 24
9b. Larger plants, 8—40 in. high; axes 600—1100 μ in diameter; dioecious 10

10a. Stipulodes 2—more per branchlet; spine-cells abundant.................. .... CHARA HORNEMANNII, p. 26
10b. Stipulodes 1 per branchlet; spine-cells rare .... CHARA BUCKELLII, p. 28

11a. Stipulodes present 12
11b. Stipulodes absent 14

12a. Stipulodes in 2 tiers ..... CHARA ZEYLANICA var. BRITTONII, p. 18
12b. Stipulodes in 1 tier 13

13a. Branchlets tipped with a corona of tiny cells ... . CHARA BRAUNII, p. 22
13b. Branchlets tipped with a single cell .... CHARA HORNEMANNII, p. 26

14a. Branchlets simple, without several-celled lateral rays .... CHARA HORNEMANNII, p. 26
14b. Branchlets either forked or with lateral rays of 2—more cells 15

15a. End segment of branchlets 1-celled 16
15b. End segment of branchlets 2—more-celled 22

16a. Heteroclemous (with 2 kinds of branchlets in whorl) .... NITELLA CLAVATA, p. 38
16b. Homoeoclemous (all branchlets in a whorl similar) 17

17a. Branchlets uniformly 1-furcate 18
17b. Branchlets occasionally 2-furcate .... NITELLA MACOUNII, p. 32
(Check also the tiny N. flexilis var. spanioclema)

18a. Fertile heads absent or, where present, not dense 19
18b. Fertile heads small, densely compact.. 21

19a. Dactyl apices blunt, acute, or apiculate, but not long acuminate .... NITELLA FLEXILIS, p. 30
19b. Dactyl apices, especially of fertile heads, long acuminate (tapering to slender point).... 20

20a. Generally without lime incrustation; gametangia on branchlets; oospore membrane smooth or granular ....NITELLA ACUMINATA, p. 34
20b. Generally with lime; gametangia foliar and at base of branchlets; oospore membrane reticulate .... NITELLA BASTINII, p. 36

21a. Very robust plants; heads with dense mucus; sterile branchlets commonly tipped with tiny corona of dactyls. . NITELLA CERNUA, p. 40

21b. Moderately large to small plants; heads without mucus; sterile branchlets generally without tiny terminal dactyls . . . . NITELLA ACUMINATA var. GREENII, p. 34

22a. End segment with blunt tip; end-cell little reduced relative to penultimate cell 23

22b. End segment sharp tipped; end-cell a much reduced cell relative to penultimate cell 25

23a. Sterile branchlets 1—more-furcate 24

23b. Sterile branchlets simple, or with lateral rays; generally lime incrusted; coarse heads resembling bird's nest . . . . TOLYPELLA NIDIFICA, p. 56

24a. Branchlets 1(—2)-furcate; without mucus . . . . . . NITELLA MACOUNII, p. 32

24b. Branchlets 2—4-furcate; heads with mucus . . . . NITELLA CAPILLATA, p. 42

25a. Branchlets monopodial (with lateral rays); heads coarse and tangled, resembling bird's nests . . . . TOLYPELLA INTRICATA, p. 58

25b. Branchlets furcate; heads not coarse 26

26a. Sterile branchlets appearing simple, but usually terminated by tiny corona of cells; heads axillary . . . . NITELLA TRANSLUCENS, p. 54

26b. Sterile branchlets furcate; heads terminal 27

27a. Heteroclemous (with 2 kinds of branchlets in whorl) . . . . NITELLA HYALINA, p. 52

27b. Homoeoclemous (branchlets in whorl uniform) 28

28a. Robust plants 6—more in. high; axes 600—more μ in diameter; dactyls occasionally small . . . . NITELLA FURCATA, p. 44

28b. Smaller plants seldom over 6 in. high; axes 200—450 μ in diameter; dactyls uniform 29

29a. Lowest branchlet node fertile; dactyls 2- or 2—3-celled; heads occasionally with mucus; oospore membrane granular or feltlike .... NITELLA GRACILIS, p. 50

29b. Lowest branchlet node sterile; dactyls 2-celled; heads without mucus; oospore membrane reticulate .... NITELLA TENUISSIMA, p. 48

SOME USEFUL REFERENCES

For other references, see lists in Wood (1952, and especially 1965).

Allen, G. O. 1954. An annotated key to the Nitelleae of North America. Bull. Torrey Bot. Club 81: 35-60.

Allen, T. F. 1888. The Characeae of America. Part 1. New York. 64 p.

Daily, Fay K. 1944. The Characeae of Nebraska. Butler Univ. Bot. St. 6: 149-171.

______. 1955. The Characeae of Indiana. Ibid. 11: 5-49.

Fritsch, F. E. 1935. The structure and reproduction of the algae. Vol. I. University Press, Cambridge. 791 p.

Groves, J., and G. R. Bullock-Webster. 1920. The British Charophyta. Vol. I. Nitelleae. Ray Society, London. 141 p.

______ and ______. 1924. The British Charophyta. Vol. II. Chareae. Ray Society, London. 129 p.

Horn af Rantzien, H. 1950. Charophyta reported from Latin America. Ark. Bot., ser. 2, 1: 355-411.

Ophel, I. L. 1952. The genus Chara in Oklahoma. Proc. Oklahoma Acad. Sci. 33: 179-182.

Prescott, G. W. 1951. Algae of the Western Great Lakes area. Cranbrook Institute of Science, Bloomfield, Michigan. 946 p.

______ 1962. Ibid., ed. 2. Brown, Dubuque. 977 p.

Robinson, C. B. 1906. The Chareae of North America. Bull. New York Bot. Gard. 4: 244-308.

Wood, R. D. 1952. The Characeae, 1951. Bot. Rev. 18: 317-353.

______. 1965. Monograph of the Characeae. *In* R. D. Wood & K. Imahori, A revision of the *Characeae*. Vol. I. Cramer, Weinheim. 904 p.

Wood, R. D., and K. Imahori. 1964. Iconograph of the Characeae. *In* R. D. Wood & K. Imahori, A revision of the *Characeae*. Vol. II. Cramer, Weinheim. 6 p. + 395 plates.

Woods, A. F. 1894. Charophyceae. p. 122-128. *In* Flora of Nebraska, 1(2). Botanical *Seminar*, Lincoln.

SOME USEFUL WORKS IN FOREIGN LANGUAGES

A large number of valuable works have been written in other tongues; for references, see bibliography in Wood (1965). See especially

Braun, A. (1867) "Die Characeen Afrika's"; Braun, A., & C. F. O. Nordstedt (1882) "Fragmente einer Monographie der Characeen"; Groves, H., and J. Groves (1911) "Characeae" p. 30-34. In I. Urban, Symbolae Antillanae, vol. 7; Migula, W. (1890-97) "Die Characeen." In L. Rabenhorst, Kryptogamen-Flora von Deutschland, Oesterreich und der Schweiz, ed. 2, vol. 5.

INDEX